Contents

Contents

Introduction

How to use this book

This book has been written to be used alongside the Edexcel AS and A2 Chemistry textbooks published by Philip Allan Updates, and written by George Facer. It aims to help you develop your study skills, make your learning more effective and give you help with your revision.

Modular courses mean that you could have examinations at different stages of your course, so you need to be prepared from the beginning.

The Edexcel specification (September 2008) divides subject matter into 24 topics. This revision book deals with each topic separately and in the order in which they occur in the specification.

The format in each topic is:

- **An introduction**
- **Things to learn** — these are mainly definitions
- **Things to understand** — includes worked examples and diagrams
- **A check list** — to check that you have covered and have understood all that is in the topic
- **Testing your knowledge and understanding** — this consists of questions that either require short answers, which are given in the margin, or those that require longer answers, which appear in the back of the book.

Hint

Formulae are used frequently rather than names of substances. A purist might object, but this is a revision book — students make mistakes more often with formulae than with names. This use of formulae will help to imprint them in your mind.

The Edexcel specification will be examined in three units for AS and three for A2. However, Units 3 and 6 are internal assessment of practical skills. Candidates will carry out a number of experiments and their work will be marked according to a mark scheme produced by Edexcel.

There are practice test papers at the end of each unit. The format and length are similar to those that you will encounter in the unit tests.

A mark scheme, together with a table of the percentages needed for each grade, is at the back of the book.

Study skills

Revision is a personal activity, and what works best for one student may not be so effective for another. However there are some golden rules:

1. Revise little and often.
2. Revise actively – do not sit and stare at your notes or this book. Write down important points or use a highlighter to mark important passages in your notes or in this book (but only if you own it!).

Hint

Do not leave your revision until the last minute. Revision should take place throughout the whole course.

3 Write down answers to the topic questions and then check them against those given.
4 Help each other. Explaining a point of chemistry to another student is a good way of clarifying your own understanding. Test each other by asking simple questions, such as formulae, definitions, and reagents and products of organic reactions.

Daily tasks

After each lesson check that your notes are complete. Try to spend 10 to 15 minutes looking through them. If there is something that you do not understand:

★ Read the relevant part in this book or your textbook and, if necessary, add to your notes so that they will be clear when you read them again.
★ Discuss the problem with another student.
★ If you still have difficulty, ask your teacher as soon as you can.
★ If you can solve the problem by yourself, your understanding may be deeper and longer lasting.

Weekly tasks

★ Go through your notes. Highlight important parts.
★ Read through the relevant parts of this book and make notes and/or highlight important points.
★ Complete any homework assignments.

End of topic tasks

When your teacher has completed a topic, you should revise that topic thoroughly. To do this:

The specification can be found on Edexcel's website, www.edexcel.com.

★ Work through your notes with a copy of the Edexcel specification.
★ Summarise your notes to the bare essentials.
★ Work through the topic in this book. Discuss any difficulties with other students.
★ Write answers to *all* the questions in the 'testing your knowledge and understanding' section of the topic.

Preparing for the unit tests

If you have followed the advice above, you will be well prepared. Since you will also be taking exams in other subjects, you should aim to start your final revision at least four weeks before the exam period.

★ Do not spend more than 30 minutes on any subject. Then switch from chemistry to another subject.
★ Take regular breaks.
★ Revise actively with pen, highlighter and paper.

When you have revised all the topics in a unit, sit down and do the relevant practice unit test. These are found at the end of each chapter. On completion, mark the test — or better still mark a friend's answers and let him or her mark yours. Then, work out where you went wrong. If you obtained low marks for a particular topic, go back to your notes and to this book and rework that topic.

Spreading revision this way over the whole course will reduce stress and will guarantee a better grade than you would obtain by leaving it all to a mad dash at the end. Chemistry is a subject in which knowledge is built up gradually. This is particularly true of organic chemistry, which many students find difficult until they suddenly get a feel for it and it all falls into place. The more thoroughly you work in the earlier stages, the easier and more enjoyable you will find the study of chemistry.

The day of the unit test

Hint

Do not take a red pen into the exam — its use is not allowed by the exam board because it will not show up when your paper is scanned ready for on-line marking.

If you have followed the advice given here, you should feel confident that you will be able to do your best. Some people find it helpful to spend a little time looking over some chemistry before going into the test. Check that you have:

- ★ two or more black pens
- ★ your calculator plus spare batteries
- ★ a watch which you should place on the desk in front of you
- ★ a ruler
- ★ a good luck charm, if that will cheer you up

Tackling the examination

- ★ Do not spend too long on a single multiple-choice question. If you are having trouble answering it, put a large circle around the question number and come back to it at the end. If you still do not have time, guess a response.
- ★ Never leave a multiple-choice answer blank. If you cannot work out the answer, guess, as there are no deductions for giving the wrong answer.
- ★ Work steadily through the paper starting at question 1.
- ★ All questions, including the multiple choice questions, are answered on the question paper itself.
- ★ If you need more room for your answer, look for space at the bottom of the page, the end of the question or after the last question.
- ★ If a question has three lines for the answer, do not write an essay. Work out the essential points that need to be made and check them against the number of marks available. The number of marks is a better guide as to how much you need write than the number of lines allotted to the question.
- ★ Do not repeat the question in your answer.
- ★ Pace yourself so that you neither run out of time nor have masses to spare at the end. If you get stuck, do not waste time. Make a note of the question number and part that caused you difficulty and continue. Later, if you have time, go back and try that part again.
- ★ Do not use correcting fluid – it is forbidden by Edexcel's regulations. Instead, neatly cross out what you have written. If, later, you realise that what you first wrote is correct, write 'ignore crossing out' beside the crossed-out work. The examiner will then mark it.

Hint

If you use these spaces, you *must* alert the examiner by adding 'continued below or continued on page XX'

Hint

As a rough guide, a question worth 15 marks should take about 15 minutes — a rate of 1 mark per minute.

Terms used in the tests

You must understand what the examiners want. Terms that are often used in the question papers are explained below:

- ★ **Define** — it is often helpful to include an example or equation to supplement your definition.
- ★ **Identify** — give the name or formula. If you write both and one is wrong, you will lose the mark.
- ★ **Formula** — make sure that the formula is unambiguous. C_3H_7OH could be propan-1-ol, $CH_3CH_2CH_2OH$, or propan-2-ol, $CH_3CH(OH)CH_3$

Hint

Formulae with methyl groups written as CH_3 are usually acceptable, especially in questions on stereo-isomerism.

- ★ **Structural formula** — it is advisable to show any double bonds.
- ★ **Displayed formula** — you must give a formula showing all the atoms and all the bonds.
- ★ **State** — no explanation is required, nor should one be given.
- ★ **Explain (or justify your answer)** — look to see how many marks there are for the question. Make sure that you give at least the same number of points in your answer.
- ★ **Deduce** — use the data supplied to answer the question.
- ★ **Hence deduce** — use the answer that you have obtained in the previous section to work out the answer.
- ★ **Suggest** — you are not expected to have learnt the answer. You should be able to apply your understanding of similar substances or situations in order to work out the answer.
- ★ **Compare** — you must make a valid comment about *both* substances.
- ★ **Calculate —** it is essential to show your working and to set your work out clearly. If you do, you can score many of the marks, even if your final answer is wrong. Give your answer to the same number of significant figures as there are in the data.
- ★ **Reagents —** full names or full formulae are required.
- ★ **Equations —** these should be balanced. Ionic equations should be used where appropriate. In organic reactions, the use of [H] and [O] to represent reducing and oxidising agents is acceptable. State symbols should be given:
  - in all thermochemical equations
  - in all electrochemical equations
  - where a precipitate or gas is produced
  - whenever they are asked for in the question
- ★ **Outline the preparation of —** only in this type of question may a flow diagram be used, with names of reagents and conditions written on the arrow.
- ★ **Stable —** this word should always be qualified, e.g. 'stable to heat', or 'the reactants are thermodynamically stable compared with the products'.

Hint

If in doubt, give your answer to 3 significant figures and for pH calculations to 2 decimal places. Do not round up in the middle of a calculation. A correct answer may not score full marks unless the working is shown.

Remember that examiners *do* try wherever possible to *give* you marks, rather than looking for ways to take them away.

Be prepared, be confident and you will do your best, which is all that anyone can ask of you.

UNIT 1

The core principles of chemistry

Topic 1.1

Formulae, equations and amount of substance

Introduction

The keys to this topic are:

- ★ to be able to calculate the number of moles from data
- ★ to set out calculations clearly

Things to learn

- ★ The Avogadro constant is the number of carbon atoms in exactly 12 g of the carbon-12 isotope. Its value is $6.02 \times 10^{23}\ mol^{-1}$.
- ★ The **relative atomic mass** of an element is the average mass of an atom of that element relative to 1/12th the mass of a carbon-12 atom.
- ★ The **relative molecular mass** of an element or compound is the average mass of a molecule (or group of ions) of that element or compound relative to 1/12th the mass of a carbon-12 atom.
- ★ One **mole** of a substance is the amount of that substance that contains 6.02×10^{23} particles of that substance. This means that one mole of a substance is its relative molecular mass expressed in grams.
- ★ The molar mass of a substance is the mass (in grams) of one mole.
- ★ Amount of substance is the number of moles of that substance.
- ★ The empirical formula is the simplest whole number ratio of the elements in the compound.

1 mol of NaOH has a mass of 40.0 g.
1 mol of O_2 has a mass of 32.0 g.

Things to understand

Calculation of empirical formulae from percentage data

This is best calculated using a table.

Element	%	% divided by r.a.m.	Divide by smallest
Carbon	48.7	48.7/12.0 = 4.1	4.1/2.7 = 1.5
Hydrogen	8.1	8.1/1.0 = 8.1	8.1/2.7 = 3
Oxygen	43.2	43.2/16.0 = 2.7	2.7/2.7 = 1

The final column gives the empirical formula. However, if any value in this column comes to a number ending in .5, .25, .33 or .67, you must multiply all the values by 2, 4 or 3 respectively to obtain integers. Here, the empirical formula is $C_3H_6O_2$.

Equations

Equations must balance. The number of atoms of an element on one side of the equation must be the same as the number of atoms of that element on the other side.

Ionic equations

Ionic equations must also balance for charge.

There are three rules:

- ★ Write the ions separately for *solutions* of ionic compounds (salts, strong acids and bases).
- ★ Write full 'molecular' formulae for solids and all covalent substances.
- ★ Spectator ions must be cancelled so that they do not appear in the final equation.

Moles

There are three ways of calculating the amount (in moles) of a substance:

- ★ For a pure substance X:

$$\textbf{amount (in moles) of X} = \frac{\textbf{mass of X (in grams)}}{\textbf{molar mass}}$$

Hint

Avoid writing mol = 0.061. Instead, state the name or formula of the substance, e.g. amount of H_2O = 0.061 mol.

Worked example

Calculate the amount of H_2O in 1.1 g of water

Answer

$1.1\ g/18.0\ g\ mol^{-1} = 0.061$ mol of H_2O

- ★ For gases:

$$\textbf{amount of gas (in moles)} = \frac{\textbf{volume (in dm}^3\textbf{)}}{\textbf{molar volume}}$$

The molar volume of a gas is $24\ dm^3\ mol^{-1}$, measured at room temperature and pressure (r.t.p.).

Worked example

Calculate the amount of H_2 in $3.2\ dm^3$ of $H_2(g)$ at r.t.p.

Answer

3.2/24 = 0.13 mol of $H_2(g)$

- ★ For solutions:

As volume in dm^3 = volume in $cm^3/1000$, the following formulae can be used:
moles = $M \times V/1000$
or
concentration = moles $\times 1000/V$
where M is the concentration in $mol\ dm^{-3}$ and V is the volume in cm^3

amount of solute (in moles) = concentration (in $\mathbf{mol\ dm^{-3}}$**)** $\times$ **volume (in** $\mathbf{dm^3}$**)**

$$\textbf{concentration} = \frac{\textbf{moles}}{\textbf{volume in dm}^3}$$

Worked example

Calculate the amount of NaOH in $22.2\ cm^3$ of a $0.100\ mol\ dm^{-3}$ solution.

Answer

$0.100 \times 0.0222 = 2.22 \times 10^{-3}$ mol of NaOH

Calculation of number of particles

The number of particles can be calculated from the number of moles:

- ★ number of molecules = moles × Avogadro constant
- ★ number of ions = moles × Avogadro constant × number of those ions in the formula

Worked example

Calculate the number of carbon dioxide molecules in 3.3 g of CO_2.

Answer

amount of CO_2 = 3.3/44.0 = 0.075 mol

number of molecules = $0.075 \times 6.02 \times 10^{23} = 4.5 \times 10^{22}$

Calculate the number of sodium ions in 5.5 g of Na_2CO_3

Answer

amount of Na_2CO_3 = 5.5/106.0 = 0.0519 mol

number of Na^+ ions = $0.0519 \times 6.02 \times 10^{23} \times 2 = 6.2 \times 10^{22}$

Calculations based on reactions

These can only be carried out if a correctly balanced equation is used.

Reacting mass questions

First, write a balanced equation for the reaction.

Then follow the route:

$$\text{mass A} \xrightarrow{\text{step 1}} \text{moles A} \xrightarrow{\text{step 2}} \text{moles B} \xrightarrow{\text{step 3}} \text{mass B}$$

For steps 1 and 3 use the relationship:

amount of A or B (in moles) = mass/molar mass

For step 2 use the stoichiometric ratio from the equation:

moles of B = moles of A × ratio B/A

Worked example

Calculate the mass of sodium hydroxide required to react with 1.23 g of silicon dioxide.

Answer

Equation: $SiO_2 + 2NaOH \rightarrow Na_2SiO_3 + H_2O$

Step 1: amount of SiO_2 = 1.23/60.1 = 0.0205 mol

Step 2: amount of NaOH = 0.0205 mol × 2/1 = 0.0410 mol

Step 3: mass of NaOH = 0.0410 × 40.0 = 1.64 g

In step 2, the stoichiometric ratio is 2:1 as there are 2NaOH to $1SiO_2$ in the equation.

Concentration of solutions

There are three formulae involved here:

$$\frac{\textbf{amount of solute (in moles)}}{\textbf{volume of solution in dm}^3} \quad \textbf{units: mol dm}^{-3}$$

$$\frac{\textbf{amount of solute (in grams)}}{\textbf{volume of solution in dm}^3} \quad \textbf{units: g dm}^{-3}$$

$$\frac{\textbf{amount of solute (in grams)}}{\textbf{1 million grams of solute}} \quad \textbf{units: ppm}$$

Parts per million (ppm) is used for measuring the concentration of material present in small quantities, such as pollutants in the air or in water supplies.

Worked example

Calculate the concentration, in $mol\ dm^{-3}$, of a solution containing 2.22 g of sodium hydroxide in 200 cm^3 of solution.

Answer

amount of NaOH = 2.22/40.0 = 0.0555 mol
concentration = 0.0555/0.200 = 0.278 $mol\ dm^{-3}$

A water sample contained 0.0025 ppm of aluminium sulfate, $Al_2(SO_4)_3$. Calculate the mass of aluminium sulfate in 1 dm^3 of water.

Answer

1 dm^3 of water has a mass of 1000 g.

mass of aluminium sulfate in 10^6 g of water = 0.0025 g
mass in 1000 g of water (1 dm^3) = $0.0025 \times 1000/10^6 = 0.0000025$ or 2.5×10^{-6} g

Gas volume calculations

1 For reactions where a gas is produced from a solid or a solution, follow the scheme below.

$$\text{mass of A} \xrightarrow{\text{step 1}} \text{moles of A} \xrightarrow{\text{step 2}} \text{moles of gas B} \xrightarrow{\text{step 3}} \text{volume of gas B}$$

For step 1 use the relationship:

$$\textbf{moles of A} = \frac{\textbf{mass of A}}{\textbf{molar mass of A}}$$

For step 2 use the stoichiometric ratio from the equation:

moles of B = moles of A × ratio of B/A

For step 3 use the relationship:

volume of gas B = moles of B × molar volume

Worked example

Calculate the volume of carbon dioxide gas evolved at r.t.p. when 7.8 g of sodium hydrogen carbonate is heated. The molar volume of a gas is 24 dm^3 at the temperature and pressure of the experiment.

Answer

In step 2, the stoichiometric ratio is 1:2 as there is $1CO_2$ to $2NaHCO_3$ in the equation.

Equation: $2NaHCO_3 \rightarrow Na_2CO_3 + H_2O + CO_2(g)$
Step 1: amount of $NaHCO_3$ = 7.8/84.0 mol = 0.09286 mol
Step 2: amount of CO_2 = 0.09286 mol × 1/2 = 0.0464 mol
Step 3: volume of CO_2 = 0.0464 mol × 24 $dm^3\ mol^{-1}$ = 1.1 dm^3

2 For calculations involving gases only, a shortcut can be used. The volumes of the two gases are in the same ratio as their stoichiometry in the equation.

Worked example

What volume of oxygen is needed to completely burn 15.6 cm^3 of ethane?

Answer

Equation: $2C_2H_6(g) + 7O_2(g) \rightarrow 4CO_2(g) + 6H_2O(l)$

Calculation: $$\frac{\text{volume of oxygen gas}}{\text{volume of ethane gas}} = \frac{7}{2} = 3.5$$

volume of oxygen (O_2) = 3.5 × 15.6 = 54.6 cm^3

Significant figures

★ You should always express your answer to the same number of significant figures as asked for in the question or as there are in the data.

Hint

If you cannot work this out in an exam, give your answer to 3 significant figures (or 2 decimal places for pH calculations), and you are unlikely to be penalised. Do not round up numbers in the middle of a calculation. Any intermediate answers should be given to at least 1 significant figure more than your final answer.

Percentage yield

First, calculate the theoretical yield from the equation using the reacting mass method as above (page 3).

Then, the % yield = $\dfrac{\text{actual yield in grams}}{\text{theoretical yield in grams}} \times 100\ \%$

The % yield is less than 100% because of competing reactions and handling losses.

Atom economy

This can be obtained from the chemical equation where:

$$\text{atom economy} = \frac{\text{mass of atoms in the desired product} \times 100}{\text{mass of atoms of all the reactants}}$$

$$= \frac{\text{molar mass of product} \times \text{number of formulae of it in the equation} \times 100}{\text{sum of molar masses of all the reactants}}$$

Worked example

Calculate the atom economy of the production of silver from silver nitrate solution and copper.

$2AgNO_3(aq) + Cu(s) \longrightarrow Cu(NO_3)_2(aq) + 2Ag$

Answer

molar masses:

$Ag = 107.9\ g\ mol^{-1}$

$AgNO_3 = 169.9\ g\ mol^{-1}$

$Cu = 63.5\ g\ mol^{-1}$

$$\text{atom economy} = \frac{107.9 \times 2 \times 100}{(169.9 \times 2) + 63.5} = \frac{21\,580}{403.3} = 53.5\%$$

Hint

Do not forget to multiply the molar masses of silver and silver nitrate by 2 as there are 2 mol of each in the equation.

Checklist

Before attempting questions on this topic, check that you can:

- ☐ calculate the empirical formula of a substance from the % composition
- ☐ write balanced ionic equations
- ☐ calculate the number of moles of a pure substance from its mass, of a solute from the volume and concentration of its solution, and of a gas from its volume
- ☐ calculate reacting masses and reacting gas volumes
- ☐ calculate the concentration of a solution in $mol\ dm^{-3}$ or ppm
- ☐ calculate the atom economy given the equation

Testing your knowledge and understanding

For the first set of questions, cover the margin, write down your answer, then check to see if you are correct.

The table below contains data that will help you.

Substance	Solubility
Nitrates	All soluble
Chlorides	All soluble, except for AgCl and $PbCl_2$
Sodium compounds	All soluble
Hydroxides	All insoluble, except for those of group 1 and barium

★ Write ionic equations for the reactions between solutions of:

a lead nitrate and potassium chloride
b magnesium chloride and sodium hydroxide
c sodium chloride and silver nitrate
d hydrochloric acid and sodium hydroxide

Answers

a $Pb^{2+}(aq) + 2Cl^-(aq) \longrightarrow PbCl_2(s)$
b $Mg^{2+}(aq) + 2OH^-(aq) \longrightarrow Mg(OH)_2(s)$
c $Cl^-(aq) + Ag^+(aq) \longrightarrow AgCl(s)$
d $H^+(aq) + OH^-(aq) \longrightarrow H_2O(l)$

★ Calculate the amount (in moles) of:

a Na in 1.23 g of sodium metal
b NaCl in 4.56 g of solid sodium chloride
c Cl_2 in 789 cm^3 of chlorine gas at room temperature and pressure
d NaCl in 32.1 cm^3 of a 0.111 mol dm^{-3} solution of sodium chloride

a 1.23/23.0 = 0.0535 mol
b 4.56/58.5 = 0.0779 mol
c 0.789/24.0 = 0.0329 mol
d $0.111 \times 0.0321 = 3.56 \times 10^{-3}$ mol

★ Calculate the volume of a 0.222 mol dm^{-3} solution of sodium hydroxide, which contains 0.0456 mol of NaOH.

0.0456/0.222 = 0.205 dm^3 = 205 cm^3

★ Calculate the number of water molecules in 1.00 g of H_2O.

$(1.00/18.0) \times 6.02 \times 10^{23} = 3.34 \times 10^{22}$

★ Calculate the number of sodium ions in 1.00 g of Na_2CO_3.

$(1.00/106.0) \times 2 \times 6.02 \times 10^{23} = 1.14 \times 10^{22}$

★ 4.44 g of solid sodium hydroxide were dissolved in water and the solution made up to 250 cm^3. Calculate the concentration in:

a g dm^{-3}
b mol dm^{-3}

a 4.44g/0.250 dm^3 = 17.8 g dm^{-3}
b 4.44/40.0 = 0.111 mol
Therefore, 0.111 mol/0.250 dm^3 = 0.444 mol dm^{-3}

The **answers** to the **numbered questions** are on page 127

1 a An organic compound contains 82.76% carbon and 17.24% hydrogen by mass. Calculate its empirical formula.
b It was found to have a relative molecular mass of 58.0. Calculate its molecular formula.

2 Balance the following equations:
a $NH_3 + O_2 \longrightarrow NO + H_2O$
b $Fe^{3+}(aq) + Sn^{2+}(aq) \longrightarrow Fe^{2+}(aq) + Sn^{4+}(aq)$

3 What mass of sodium hydroxide is needed to react with 2.34 g of phosphoric(V) acid, H_3PO_4, to form the salt Na_3PO_4?

4 When sodium chloride solution is added to lead nitrate solution, a precipitate of lead chloride is formed.

$$2NaCl(aq) + Pb(NO_3)_2(aq) \longrightarrow PbCl_2(s) + 2NaNO_3(aq)$$

a Calculate the atom economy for the preparation of lead chloride by this method.
b If hydrochloric acid was used instead of sodium chloride, would the atom economy be higher or lower? Justify your answer.

5 At r.t.p., what volume of hydrogen sulfide gas, H_2S, is required to react with $25\,cm^3$ of a $0.55\,mol\,dm^{-3}$ solution of bismuth nitrate, $Bi(NO_3)_3$? The molar volume of a gas is $24\,dm^3\,mol^{-1}$ under these conditions. The reagents react according to the equation:

$$3H_2S(g) + 2Bi(NO_3)_3(aq) \rightarrow Bi_2S_3(s) + 6HNO_3(aq)$$

6 What volume of hydrogen gas is produced by the reaction of $33\,dm^3$ of methane gas with steam, according to the equation:

$$CH_4(g) + H_2O(g) \rightarrow CO(g) + 3H_2(g)?$$

7 Calculate the number of aluminium ions in $1.0\,cm^3$ of:

a a 0.030 ppm solution of aluminium sulfate, $Al_2(SO_4)_3(aq)$

b a $0.030\,nmol\,dm^{-3}$ solution of aluminium sulfate

(molar mass of aluminium sulfate = $342.3\,g\,mol^{-1}$; Avogadro constant = $6.02 \times 10^{23}\,mol^{-1}$; 1 nmol (nanomole) = 10^{-9} mol)

8 Calculate the mass of sodium sulfate, Na_2SO_4, produced when 4.0 g of sodium hydroxide is mixed with a solution containing 5.0 g of sulfuric acid:

$$2NaOH + H_2SO_4 \rightarrow Na_2SO_4 + 2H_2O$$

Topic 1.2

Energetics

Introduction

- State symbols should always be used in equations in this topic.
- In definitions in this topic, there are three key points:
  - the enthalpy change per mole
  - the substance the mole refers to (for example, 1 mol of substance being formed/1 mol of atoms being produced/1 mol of substance being burnt).
  - the standard conditions
- It always helps to add an equation with state symbols as an example because this may gain marks lost by an omission in the definition.

Things to learn

- An **exothermic reaction** produces heat, so that heat is then given out to the surroundings. For all exothermic reactions, ΔH is *negative* (see Figure 1.1). This means that chemical energy is being converted into thermal (heat) energy.
- An **endothermic reaction** loses heat, so that heat is then taken in from the surroundings. For all endothermic reactions, ΔH is *positive* (see Figure 1.2).

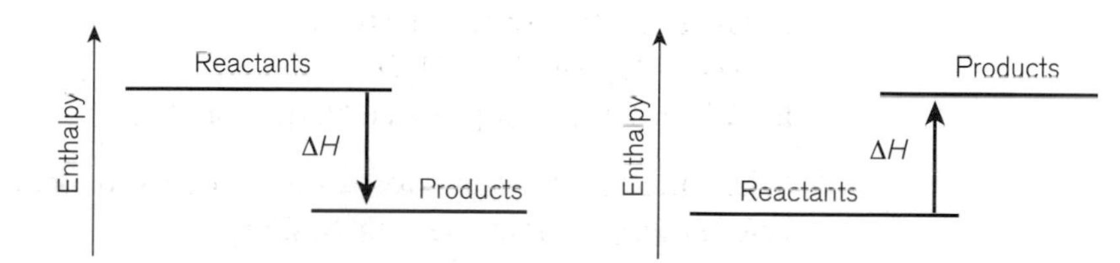

Figure 1.1 Enthalpy diagram for an exothermic reaction

Figure 1.2 Enthalpy diagram for an endothermic reaction

- Standard conditions are:
  - a pressure of 1 atmosphere
  - a stated temperature (usually 298 K)
  - solutions, if any, at a concentration of $1.00\,mol\,dm^{-3}$

- substances in their most stable states, e.g. carbon as graphite not diamond, water at 298K as a liquid.

★ Standard **enthalpy of reaction**, $\Delta H_r^\ominus$, is the enthalpy change when the molar quantities as written in the equation react at 1 atm pressure and a stated temperature (usually 298 K). For example, $\Delta H_r^\ominus$ for sulfur dioxide reacting with oxygen according to the equation:

$2SO_2(g) + O_2(g) \longrightarrow 2SO_3(g)$

is for 2 mol of SO_2 gas reacting with 1 mol O_2 gas to form 2 mol of SO_3 gas.

It follows from this definition that the enthalpy of formation of an element in its stable state is zero.

★ Standard **enthalpy of formation**, $\Delta H_f^\ominus$, is the enthalpy change when *one mole* of a compound is formed from its *elements* in their most stable states at 1 atm pressure and a stated temperature (usually 298 K). For example, $\Delta H_f^\ominus$ for ethanol is the enthalpy change for the reaction:

$2C(graphite) + 3H_2(g) + \frac{1}{2}O_2(g) \longrightarrow C_2H_5OH(l)$

★ Standard **enthalpy of combustion**, $\Delta H_c^\ominus$, is the enthalpy change when *one mole* of a substance is completely burned in oxygen at 1 atm pressure and a stated temperature (usually 298 K). For example, $\Delta H_c^\ominus$ for ethane is the enthalpy change for the reaction:

$C_2H_6(g) + 3\frac{1}{2}O_2(g) \longrightarrow 2CO_2(g) + 3H_2O(l)$

★ Standard **enthalpy of atomisation** of an element, $\Delta H_a^\ominus$, is the enthalpy change when *one mole* of a *gaseous atom* is formed from the element in its stable state at 1 atm pressure and a stated temperature (usually 298 K). For example, $\Delta H_a^\ominus$ for bromine is the enthalpy change for the reaction:

$\frac{1}{2}Br_2(l) \longrightarrow Br(g)$

Hint

You should always give an equation as an example for this and other enthalpy definitions.

★ Standard **enthalpy of neutralisation** $\Delta H^\ominus_{neut}$ of an acid is the enthalpy change when the acid is neutralised by a base and *one mole of water* is produced at 1 atm pressure and a stated temperature (usually 298 K). For example, $\Delta H^\ominus_{neut}$ for sulfuric acid, with sodium hydroxide solution, is the enthalpy change for the reaction:

$\frac{1}{2}H_2SO_4(aq) + NaOH(aq) \longrightarrow \frac{1}{2}Na_2SO_4(aq) + H_2O(l)$

For hydrochloric acid, it is for the reaction:

$HCl(aq) + NaOH(aq) \longrightarrow NaCl(aq) + H_2O(l)$

For any strong acid being neutralised by any strong base, it is for the reaction:

$H^+(aq) + OH^-(aq) \longrightarrow H_2O(l) \quad \Delta H^\ominus_{neut} = -57\ kJ\ mol^{-1}$

★ **Average bond enthalpy** is the *average* enthalpy change when *one mole* of that bond is broken into *separate gaseous atoms* of the elements concerned (at 1 atm pressure). It is *always* endothermic.

★ **Hess's law** states that the enthalpy change for a given reaction is independent of the route by which the reaction takes place, provided the states of the reactants and products are the same in both routes (see Figure 1.3). Thus, the enthalpy change proceeding *directly* from reactants to products is the same as the *sum* of the enthalpy changes of all the reactions for the change to be carried out in two or more steps.

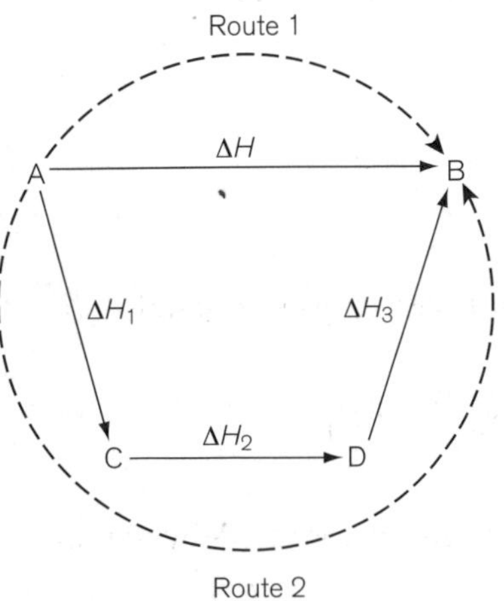

Figure 1.3 A reaction cycle

Things to understand

Calculation of ΔH_r from ΔH_f data

★ You can use the expression:

$\Delta H_{reaction}$ = **the sum of $\Delta H_{formation}$ of products – the sum of $\Delta H_{formation}$ of reactants**

★ Remember that if you have two moles of a substance, you must double the value of ΔH_f.

★ The enthalpy of formation of an element (in its stable state) is zero.

> **Hint**
> Never say
> ΔH = products – reactants
> as it will give the wrong sign if combustion or bond enthalpy data are used.

Worked example

Calculate the standard enthalpy change for the reaction:

$$4NH_3(g) + 5O_2(g) \longrightarrow 4NO(g) + 6H_2O(g)$$

Relevant $\Delta H_f^\ominus$ data (kJ mol^{-1}): $NH_3(g)$ –46.2; NO(g) +90.4; $H_2O(g)$ –242

Answer

$\Delta H_r^\ominus = \sum\Delta H_f^\ominus(\text{products}) - \sum\Delta H_f^\ominus(\text{reactants})$

$= 4 \times (+90.4) + 6 \times (-242) - 4 \times (-46.2) - 5 \times (0) = -906\ \text{kJ mol}^{-1}$

Calculation of ΔH_f from ΔH_c data

To do this you use the alternative route:

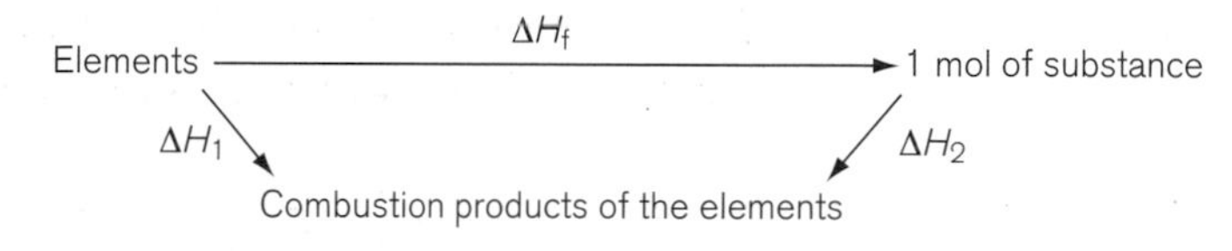

where ΔH_1 is the sum of the enthalpies of combustion of the elements, taking into account the number of moles of each element

and ΔH_2 is the enthalpy of combustion of 1 mol of the substance

> If an equation is reversed, the sign of its ΔH must be changed, so $\Delta H_f = \Delta H_1 - \Delta H_2$

Worked example

Calculate the standard enthalpy of formation of propan-1-ol, $C_2H_5CH_2OH$. Relevant $\Delta H_c^\ominus$data (kJ mol^{-1}): propan-1-ol(l) –2010; C(graphite) –394; $H_2(g)$ –286

Answer

$3C(s) + 4H_2(g) + \frac{1}{2}O_2(g) \longrightarrow C_2H_5CH_2OH(l)$

$3 \times \Delta H_c(\text{graphite})$ $\quad 4 \times \Delta H_c(H_2)$ $\quad \Delta H_c(C_2H_5CH_2OH)$

$3CO_2(g) + 4H_2O(l)$

$\Delta H_f^\ominus = \sum\Delta H_c^\ominus(\text{elements}) - \Delta H_c^\ominus(\text{ compound})$

$= 3 \times (-394) + 4 \times (-286) - (-2010) = -316\ \text{kJ mol}^{-1}$

Use of Hess's Law to calculate ΔH for a reaction that cannot be measured directly

The enthalpy change for the reaction:

$$CuSO_4.5H_2O(s) \longrightarrow CuSO_4(s) + 5H_2O(l)$$

cannot be measured directly.

Hess's law can be used to find the value by measuring the enthalpies of solution of hydrated and anhydrous copper(II) sulfate.

Equation 1: $CuSO_4.5H_2O(s) + aq \longrightarrow CuSO_4(aq) + 5H_2O(l) \qquad \Delta H_1$
Equation 2: $CuSO_4(s) + aq \longrightarrow CuSO_4(aq) \qquad \Delta H_2$

If equation 2 is reversed and added to equation 1, the equation for the dehydration of hydrated copper(II) sulfate is obtained.

ΔH for the dehydration of hydrated copper(II) sulfate = $\Delta H_1 + (-\Delta H_2)$

Calculation of ΔH values from laboratory data

★ Calculation of heat change from temperature change:

The mass in this expression is the mass of the solution, not the mass of the reactants.

heat change = mass × specific heat capacity × change in temperature

★ Then heat change per mole = $\dfrac{\text{heat change in experiment}}{\text{the number of moles reacted}}$

★ If the temperature increases, the reaction is exothermic. The enthalpy change is a *negative* number, therefore ΔH = –heat change per mole

★ If the temperature decreases, the reaction is endothermic. The enthalpy change is a *positive* number, therefore ΔH = +heat change per mole.

Worked example

A 0.120 g sample of ethanol was burnt and the heat produced warmed 250 g of water from 17.30°C to 20.72°C. The specific heat capacity of water is 4.18 J g^{-1} °C^{-1} Calculate ΔH_c of ethanol.

Answer

heat change = 250 g × 4.18 J g^{-1} °C^{-1} × 3.42°C^{-1} = 3574 J

moles of ethanol = $\dfrac{0.120}{46}$ = 0.00261

heat change per mole = 3574 J/0.00261 mol = 1370 × 10^3 J mol^{-1}/0.00261 mol

ΔH_c = –1370 kJ mol^{-1}

Since the temperature rose, the reaction is exothermic. Therefore, ΔH is negative.

Calculation of ΔH values from average bond enthalpies

★ Draw structural formulae for all the reactants and all the products.

★ Decide what bonds are broken in the reaction and calculate the energy required for this (endothermic).

★ Decide what bonds are formed in the reaction and calculate the energy released by this (exothermic).

★ Add the positive bond-breaking enthalpy to the negative bond-making enthalpy.

Worked example

Calculate the enthalpy of the reaction:

$$H_2C{=}CH_2 + H{-}H \rightarrow H_3C{-}CH_3$$

Relevant average bond enthalpy data (kJ mol^{-1}): C–C +348; C=C +612; H–H +436; C–H +412

Answer

Break (endo)		Make (exo)	
C=C	+612	C–C	–348
H–H	+436	2 × C–H	–824
	+1048		–1172

$\Delta H_{reaction}$ = +1048 + (–1172) = –124 kJ mol^{-1}

Checklist

Before attempting questions on this topic, check that you:

- ☐ know the standard conditions for enthalpy changes
- ☐ know the signs of ΔH for an exothermic and an endothermic reaction
- ☐ can define the standard enthalpies of reaction, formation, combustion, atomisation and neutralisation
- ☐ understand an energy level diagram
- ☐ can define and use Hess's law
- ☐ know how to calculate values of ΔH from laboratory data
- ☐ can use bond enthalpies to calculate enthalpies of reaction

Testing your knowledge and understanding

Answers

a $-395\ \text{kJ mol}^{-1}$
b $+790\ \text{kJ mol}^{-1}$

$25 \times 4.2 \times 5.0 = 525\ \text{J}$

The mass used in the calculation is the mass of the solution (25 g), not the mass of reactant (2 g).

The **answers** to the **numbered questions** are on page 127

For the following set of questions, cover the margin, write your answers, then check to see if you are correct.

★ $\Delta H_r^\ominus$ for $2SO_2(g) + O_2(g) \rightarrow 2SO_3(g)$ is $-790\ \text{kJ mol}^{-1}$
What is the value of $\Delta H_r^\ominus$ for the following reactions?
- a $SO_2(g) + \frac{1}{2}O_2(g) \rightarrow SO_3(g)$
- b $2SO_3(g) \rightarrow 2SO_2(g) + O_2(g)$

★ 2.0 g of a metal oxide was added to 25 g of a solution of hydrochloric acid. The temperature rose by 5.0°C and the specific heat capacity of the solution is $4.2\ \text{J K}^{-1}\ \text{mol}^{-1}$. What is the value of the heat change?

1 State the conditions used when measuring ***standard*** enthalpy changes.

2 Give equations, with state symbols, to represent the following enthalpy changes:
- a the enthalpy of formation of ethanoic acid, $CH_3COOH(l)$
- b the enthalpy of combustion of ethanoic acid.
- c the enthalpy of neutralisation of an aqueous solution of ethanoic acid with aqueous sodium hydroxide

3 Draw an enthalpy level diagram for the following sequence of reactions:

$\frac{1}{2}N_2(g) + \frac{1}{2}O_2(g) \rightarrow NO(g)$ $\quad \Delta H = +90.3\ \text{kJ mol}^{-1}$

$NO(g) + \frac{1}{2}O_2(g) \rightarrow NO_2(g)$ $\quad \Delta H = -57.1\ \text{kJ mol}^{-1}$

Hence calculate the enthalpy change for the reaction:

$\frac{1}{2}N_2(g) + O_2(g) \rightarrow NO_2(g)$

4 Given that the standard enthalpies of formation of $NO_2(g)$ and $N_2O_4(g)$ are $+33.9\ \text{kJ mol}^{-1}$ and $+9.7\ \text{kJ mol}^{-1}$ respectively, calculate the enthalpy change for the reaction:

$2NO_2(g) \rightarrow N_2O_4(g)$

5 The standard enthalpy of combustion of lauric acid, $CH_3(CH_2)_{10}COOH(s)$, which is found in some animal fats, is $-7377\ \text{kJ mol}^{-1}$.
The standard enthalpies of combustion of C(s) and $H_2(g)$ are $-394\ \text{kJ mol}^{-1}$ and $-286\ \text{kJ mol}^{-1}$ respectively.
Calculate the standard enthalpy of formation of lauric acid.

6 100 cm^3 of 1.00 mol dm^{-3} HCl was added to 100 cm^3 of 1.00 mol dm^{-3} NaOH in a polystyrene cup. The temperature of each solution was initially 21.10°C. On mixing, the temperature rose to 27.90°C. Determine the enthalpy of neutralisation. You may assume that the polystyrene cup has a negligible heat capacity, the solution has a density of 1.00 g cm^{-3} and that the final solution has a specific heat capacity of 4.18 J g^{-1} $°C^{-1}$.

7 The enthalpy change for the reaction:

$$2NaHCO_3(s) \rightarrow Na_2CO_3(s) + H_2O(l) + CO_2(g)$$

cannot be measured directly. Given samples of solid sodium carbonate and sodium hydrogencarbonate and a supply of dilute hydrochloric acid, plan an experiment that enables the enthalpy change for the decomposition to be measured using Hess's law.

8 Calculate the enthalpy change for the reaction:

$$C_2H_4(g) + H_2O(g) \rightarrow CH_3CH_2OH(g)$$

Relevant average bond enthalpy data (kJ mol^{-1}): C–C +348; C=C +612; C–H +412; H–O +463; C–O +360

Topic 1.3 Atomic structure and the periodic table

Introduction

- ⋆ The electrons in an atom are in orbits (or shells) around the positive nucleus.
- ⋆ The orbits are divided into suborbits (orbitals) known as the *s*-, *p*- *d*- (and *f*-) orbitals
- ⋆ The periodic table is divided into blocks:
  - *s*-block — groups 1 and 2
  - *p*-block — groups 3, 4, 5, 6, 7 and 0
  - *d*-block — the elements between the *s*- and *p*-blocks
- ⋆ The energy required to remove an electron from an atom depends on the pull of the nucleus on the electron (the effective nuclear charge, which equals nuclear charge less the electron shielding) and the distance it is from the nucleus (which orbital it is in).

Things to learn

- ⋆ **Atomic number (*Z*)** of an element is the number of protons in the nucleus of the atom.
- ⋆ **Mass number** of an isotope is the number of protons plus the number of neutrons in the nucleus.
- ⋆ **Isotopes** are atoms of the same element that have the same number of protons but different numbers of neutrons. They have the same atomic number but different mass numbers.
- ⋆ **Relative atomic mass (A_r)** of an element is the *average* mass (taking into account the abundance of each isotope) of the atoms of that element relative to 1/12 the mass of a carbon-12 atom.
- ⋆ **Relative isotopic mass** is the mass of one isotope relative to 1/12 the mass of a carbon-12 atom.
- ⋆ **Relative molecular mass (M_r)** of a substance is the sum of all the relative atomic masses of its constituent atoms.

This is sometimes called the relative formula mass (particularly for ionic substances).

Ionisation energies are always endothermic and relate to the formation of a positive ion.

- **Molar mass** is the mass of one mole of a substance. The units are $g\,mol^{-1}$ and it is numerically equal to the relative molecular mass.
- **First ionisation energy** is the amount of energy required per mole to *remove* one electron from the outer orbit of *each gaseous atom* to form a singly charged positive ion:
  $X(g) \rightarrow X^+(g) + e^-$
- **Second ionisation energy** is the energy change per mole for the removal of an electron from a singly charged positive gaseous ion to form a doubly charged positive ion:
  $X^+(g) \rightarrow X^{2+}(g) + e^-$
- **First electron affinity** is the energy change per mole for the addition of one electron to a gaseous atom to form a singly charged negative ion:
  $X(g) + e^- \rightarrow X^-(g)$
- **Second electron affinity** is the energy change per mole for the addition of an electron to a singly charged negative gaseous ion. It is always endothermic:
  $X^-(g) + e^- \rightarrow X^{2-}(g)$
- *s*-block elements are those where the *s*-orbital is being filled. They are in groups 1 and 2.
  Similar definitions apply to *p*-block (groups 3 to 7 and 0) and *d*-block (Sc to Zn) elements.

Things to understand

Mass spectra

- An element or compound is first vaporised and then bombarded with high-energy electrons. These remove an electron from the atom or molecule to form a **positive ion**. This ion is then *accelerated* by an electric potential, *deflected* according to its mass by a magnetic field and finally *detected*.
- Metals and the noble gases form singly charged positive ions in the ratio of the abundance of their isotopes.
- Non-metals also give molecular ions. For example Br_2, which has two isotopes ^{79}Br (50%) and ^{81}Br (50%), gives three lines at *m/e* values of 158, 160 and 162 in the ratio 1:2:1 (as well as lines at *m/e* = 79 and 81). These are caused by $(^{79}Br{-}^{79}Br)^+$, $(^{79}Br{-}^{81}Br)^+$ and $(^{81}Br{-}^{81}Br)^+$
- The relative atomic mass of an element can be calculated from mass spectral data as follows:

A_r = the sum of (mass of each isotope × % of that isotope)/100

Hint

A common error is to miss out the + charge on the formula of a species responsible for a line in a mass spectrum.

Worked example

Boron was analysed in a mass spectrometer.

Calculate the relative atomic mass of boron using the results below.

m/e value	% abundance
10	18.7
11	81.3

Answer

$A_r = ((10 \times 18.7) + (11 \times 81.3))/100 = 10.8$

- The spectrum of an organic molecule, for example ethanol, CH_3CH_2OH, contains peaks due to the molecular ion $(CH_3CH_2OH)^+$ (*m/e* = 46) and peaks due to fragments, such as $(CH_3)^+$ at *m/e* = 15 and $(CH_2OH)^+$ at *m/e* = 46 – 15 = 31.

- Modern uses of mass spectrometry include:
  - identifying drugs in urine samples of athletes
  - estimating the age of rocks by analysing the ratio of a radioactive element, for example ^{40}K, to its daughter product, in this case, ^{40}Ca.

Electron structure

- The first shell has only an *s*-orbital.
- The second shell has one *s*-and three *p*-orbitals.
- The third and subsequent shells have one *s*-, three *p*- and five *d*-orbitals.
- Each orbital can hold a maximum of *two* electrons
- The shape of an *s*-orbital is: and that of a *p*-orbital is:

- The order of filling orbitals is shown in Figure 1.4.

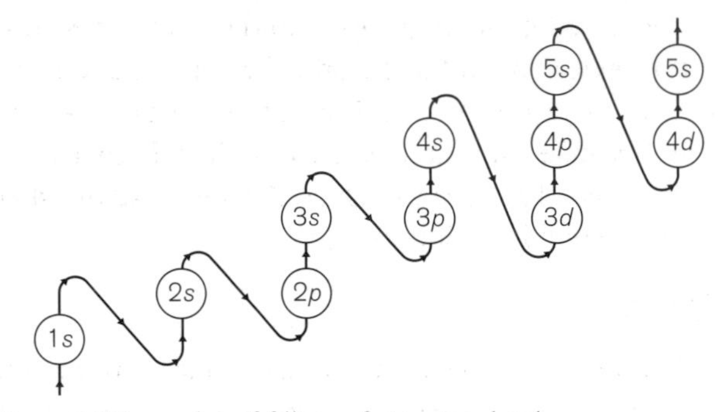

Figure 1.4 The order of filling of atomic orbitals

Electron configurations can be shown in two ways:

- The *s*-, *p*-, *d*-notation. For chlorine (atomic number 17), this is:
  $1s^2\ 2s^2\ 2p^6\ 3s^2\ 3p^5$
  This can be abbreviated to: [Ar] $3s^2\ 3p^5$ where [Ar] represents the electronic configuration of argon.
- The electrons in a box notation. For phosphorus (Z = 15), this is:

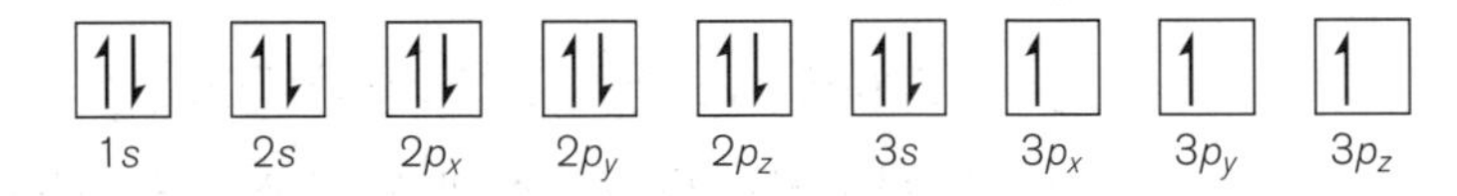

Sizes of atoms and ions

- Atoms become *smaller* going across a period from left to right. This is because the nuclear charge increases, pulling the electrons closer to the nucleus, though the number of shells is the same.
- Atoms become *larger* going down a group because there are more shells of electrons.
- A positive ion is smaller than the neutral atom from which it was formed because the ion has one electron fewer than the atom.
- A negative ion is larger than the neutral atom because the increased repulsion between the electrons causes them to spread out.

In an atom, the outer electrons are shielded from the pull of the nucleus by the electrons in shells nearer to the nucleus (the inner electrons).

First ionisation energy

★ There is a general increase going from left to right across a period (see Figure 1.5). This is caused mainly by the increased nuclear charge (atomic number) without an increase in the number of inner shielding electrons.

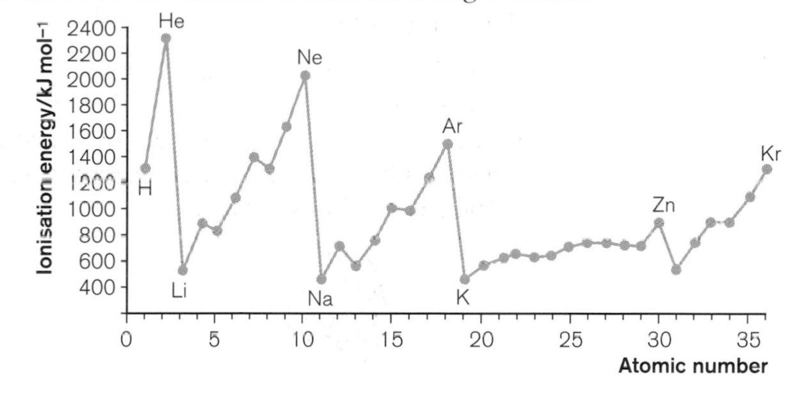

Figure 1.5 First ionisation energies (kJ mol^{-1}) of the elements up to krypton

★ The reason for the slight decrease between groups 2 and 3 is that in the group 3 element it is easier to remove an electron from the higher energy *p*-orbital. The decrease between groups 5 and 6 occurs because, in group 6, the repulsion of the two electrons in the p_x-orbital makes it easier to remove one of them.

★ There is a decrease going down a group — for example, Li to Na to K etc. This is because the outer electron is in a higher energy orbit that is further from the nucleus. (The extra nuclear charge is balanced by the same increased number of inner shielding electrons.)

Hint

You should be able to sketch the variation of first ionisation energies with atomic number for the first 20 elements.

Periodicity

★ The regular pattern of the first ionisation energies is an example of **periodicity** — low value at group 1, increasing to a peak at group 0 with slight dips at groups 3 and 6. It also shows that after each noble gas, a new electron shell starts to fill.

★ The melting points of the elements in periods 2 and 3 show a similar pattern. The melting point of the metals increase as the number of valence electrons increases from 1 to 3. Carbon has a giant structure with a very high melting point. This is followed by low melting points for the remaining non-metals — their values depend on the complexity of the molecules. The pattern is repeated in period 3 (Na to Ar).

Successive ionisation energies

★ The second ionisation energy of an element is always greater than the first because the second electron is removed from a positive ion.

★ When there is a very big jump in the value of successive ionisation energies an electron is being removed from a lower shell — for example, if this jump happens from the fourth to the fifth ionisation energy, four electrons have been removed from the outer shell during the first four ionisations, and so the element is in group 4.

The first six ionisation energies (kJ mol^{-1}) of a particular element are:

1st	786
2nd	1580
3rd	3230
4th	4360
5th	16000
6th	20000

There is a big jump after the fourth ionisation, so the element is in group 4.

Electron affinity

★ The first electron affinity value is always negative (exothermic) because a negative electron is being brought towards the positive nucleus in a neutral atom. First electron affinities are most exothermic for the halogens.

★ The second electron affinity value is always positive (endothermic) because a negative electron is being added to a negative ion.

Checklist

Before attempting the questions on this topic, check that you can:

- ☐ define A_r, M_r and relative isotopic mass
- ☐ calculate the number of neutrons in an isotope, given the atomic and mass numbers
- ☐ outline how a mass spectrometer works and give an example of a modern use of a mass spectrometer
- ☐ calculate the relative atomic mass of an element from mass spectral data
- ☐ define first and subsequent ionisation energies
- ☐ explain the changes in first ionisation energies for elements across a period and down a group
- ☐ give examples of periodicity
- ☐ deduce an element's group from successive ionisation energies
- ☐ work out the electronic structure of the first 36 elements
- ☐ sketch the shape of *s*- and *p*-orbitals
- ☐ define first and second electron affinities

Testing your knowledge and understanding

For the following set of questions, cover the margin, write your answers, then check to see if you are correct. (You may refer to a periodic table.)

★ State the masses and charges (relative to a neutron) of the following:
  a a proton
  b a neutron
  c an electron

★ How many neutrons are there in an atom of $^{23}_{11}Na$?

★ What is the atomic number of an element that has an atom of mass number 99 and which contains 56 neutrons ?

★ Write equations with state symbols for:
  a the first ionisation of sodium
  b the first ionisation of chlorine
  c the second ionisation of magnesium

★ Using the 1*s* 2*s* 2*p* etc. notation, give the electronic structure of the elements:
  a $_{6}C$
  b $_{12}Mg$
  c $_{26}Fe$

1 Explain the difference between relative isotopic mass and relative atomic mass. Illustrate your answer with reference to a specific element.

2 Lithium has naturally occurring isotopes of mass numbers 6 and 7. Explain why its relative atomic mass is 6.9 not 6.5.

Answers

a mass 1, charge +1
b mass 1, charge 0
c mass 1/1860, charge –1

23 – 11 protons = 12 neutrons

99 – 56 neutrons = 43

a $Na(g) \longrightarrow Na^+(g) + e^-$
b $Cl(g) \longrightarrow Cl^+(g) + e^-$
c $Mg^+(g) \longrightarrow Mg^{2+}(g) + e^-$

a $1s^2\ 2s^2\ 2p^2$
b $1s^2\ 2s^2\ 2p^6\ 3s^2$.
c $1s^2\ 2s^2\ 2p^6\ 3s^2\ 3p^6\ 3d^6\ 4s^2$

The **answers** to the **numbered questions** are on page 128

3 Magnesium was analysed in a mass spectrometer. Peaks were found at three different *m/e* ratios.

m/e	% abundance
24	78.6
25	10.1
26	11.3

Calculate the relative atomic mass of magnesium.

4 Gallium has a relative atomic mass of 69.8. Its mass spectrum shows two peaks at *m/e* values of 69 and 71.
Calculate the percentage of each isotope in gallium.

5 Draw diagrams to represent 1*s*-, 2*s*- and 2*p*-orbitals.

6 The successive ionisation energies of an element X are given in the table below.

Ionisation energy	1st	2nd	3rd	4th	5th	6th	7th	8th
Value/kJ mol^{-1}	1060	1900	2920	4960	6280	21 200	25 900	30 500

In which group does element X occur? Give your reasons.

7 a For the first 11 elements, sketch a graph of first ionisation energy against atomic number.

b Explain why the ionisation energy is lower for:

(i) the element of atomic number $Z = 5$ than that for the element of $Z = 4$

(ii) the element of $Z = 8$ than that for the element of $Z = 7$

(iii) the element of $Z = 11$ than that for the element of $Z = 3$

8 Using the electron in a box notation, give the electron structure of chlorine ($Z = 17$).

9 Explain, in terms of electron structure, why the chemical properties of lithium, sodium and potassium are similar but not identical.

10 a Give the equation, with state symbols, that represents the first electron affinity of:

(i) lithium

(ii) chlorine

(iii) oxygen

b Give the equation that represents the second electron affinity of oxygen.

c Why is the second electron affinity of oxygen endothermic whereas the first electron affinities of lithium, chlorine and oxygen are all exothermic?

Topic 1.4 Bonding

Introduction

★ Much of chemistry depends upon Coulomb's law which states that the electrostatic force of attraction, *F*, is given by:

$$F \propto \frac{(q_+ q_-)}{r^2}$$

where q_+ and q_- are the charges on the objects (the nucleus, an electron, an ion etc.) and r^2 is the square of the distance between their *centres*. This means that the greater the charge, the stronger the force, and the further the centres are apart, the weaker the force.

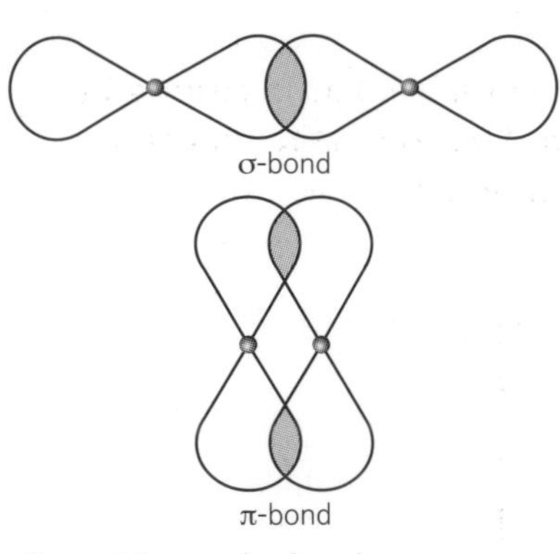

Figure 1.6 σ- and π-bonds

Things to learn

★ An **ionic bond** is the net force of attraction between a positive ion (cation) and a negative ion (anion). A cation is formed when an atom loses one or more electrons. An anion is formed when an atom gains one or more electrons.

★ A **covalent bond** is the attraction of two nuclei for a pair of shared electrons. It results from the overlap of an orbital containing one electron in one atom and an orbital that contains one electron belonging to the other atom. The overlap can be head on, which results in a σ-bond or, side by side, which results in a π-bond. A double bond is a σ-bond and a π-bond with two pairs of electrons being shared.

★ A **dative covalent bond** is a covalent bond formed when one of the overlapping orbitals contained two electrons and the other none.

> In hydrated cations, such as $[Mg(H_2O)_6]^{2+}$, the water molecules are bonded to the magnesium ion by dative covalent bonds. The lone pairs on the oxygen atoms bond with the empty *s*- and *p*-orbitals in the Mg^{2+} ion.

★ A **metallic bond** results from the delocalisation of the outer electrons over all the atoms. It can be regarded as the attraction between the 'sea' of electrons and the positive ions, which are in a lattice.

★ The **lattice energy**, $\Delta H_{\text{lattice}}$, of an ionic solid is the energy change when *one mole* of an ionic solid is formed from its *gaseous ions*.

Things to understand

Ionic bonding

★ An ionic crystal consists of a giant lattice of alternating cations and anions.

★ The force between ions in an ionic crystal depends on the sizes and charges on the ions.

★ The larger the charge, the stronger is the force of attraction.

★ The smaller the *sum* of the ionic radii, the stronger is the force of attraction.

★ The ionic radius increases down a group of the periodic table as the number of electron shells increases — so, $Li^+ < Na^+ < K^+$.

★ The ions N^{3-}, O^{2-}, F^-, Na^+, Mg^{2+} and Al^{3+} are isoelectronic; that is they all have the same electron structure (2,8). However, negative ions are larger than positive ions and increase in size as the charge increases. Positive ions decrease in radius as the charge increases. So, the radius decreases in the order $N^{3-} > O^{2-} > F^- > Na^+ > Mg^{2+} > Al^{3+}$.

Evidence for ions

Molten or aqueous solutions conduct electricity by the movement of ions. If some green copper chromate is placed on a strip of wet filter paper and the ends of the paper are connected to a power supply, the blue copper ions move towards the negative terminal and the yellow chromate ions move towards the positive terminal.

'Dot-and-cross' diagrams

The cations of group 1, 2 and 3 metals and the anions of elements in groups 5, 6 and 7 have the electron configuration of a noble gas. Ions of the *d*-block metals do not have noble gas electron configurations.

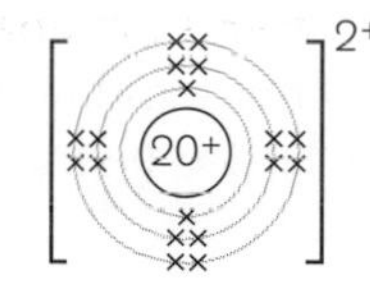

Figure 1.7 Dot-and-cross diagram for a calcium ion

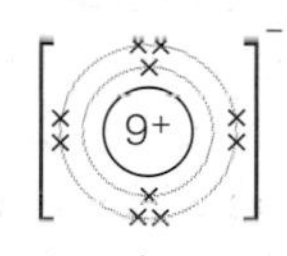

Figure 1.8 Dot-and-cross diagram for a fluoride ion

The dot-and cross-diagrams for a calcium ion and a fluoride ion are shown in Figures 1.7 and 1.8 respectively.

Born–Haber cycles

The reaction between a metal and a non-metal can be considered as occurring in several steps. The enthalpy change for the direct reaction, ΔH_f, equals the sum of the enthalpy changes of all the steps. This is shown in a Born–Haber cycle.

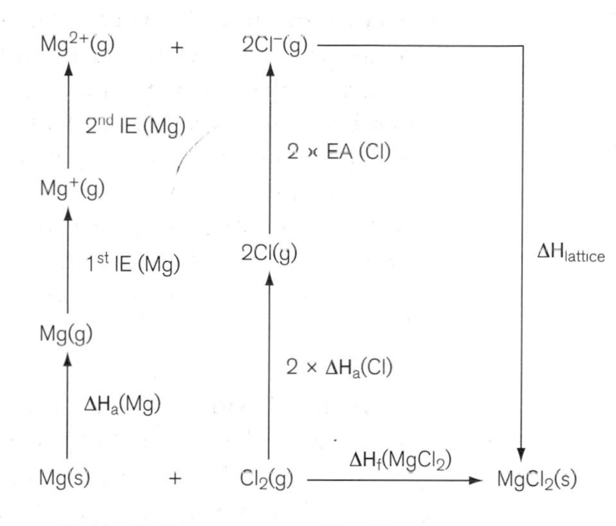

Figure 1.9 The Born–Haber cycle for magnesium chloride

ΔH_a is the enthalpy of atomisation (see p. 8).
IE is the ionisation energy.
EA is the electron affinity (which is always per mole of gaseous ions).

> Lattice energies are always exothermic.

$\Delta H_{lattice}$ is the **lattice energy** of magnesium chloride. It is defined as the energy change when *one mole* of an ionic solid is formed from its *gaseous ions*.

> The lattice energy can be evaluated from the experimental values in the Born-Haber cycle. This value is called the *experimental* lattice energy. It can also be calculated assuming that the solid is 100% ionic. This is called the *theoretical* lattice energy.

$$\Delta H_f = \Delta H_a(\text{Mg}) + 1^{st}\text{IE(Mg)} + 2^{nd}\text{IE(Mg)} + 2 \times \Delta H_a(\text{Cl}) + 2 \times \text{EA(Cl)} + \Delta H_{lattice}$$

Covalence in ionic solids

★ If the experimental value for lattice energy differs significantly from the theoretical value, the solid has some covalent character. The greater the difference, the greater is the extent of covalency.

★ The extent of covalency is determined by the extent to which the cation **polarises** the anion.

★ Cations with a *small* radius and/or high charge have a large charge density, and so are very polarising. Anions with a *large* radius and/or high charge are very polarisable. If either the cation is very polarising or the anion is very polarisable, the outer electrons in the anion are pulled towards the cation and the bond has some covalent character.

Formulae of metal ions

> **Hint**
> The reason that sodium chloride is NaCl and not $NaCl_2$ is not due to some imagined 'stability' of a noble gas electron configuration.

★ NaCl or $NaCl_2$? In order to form $NaCl_2$, a second electron would have to be removed from the sodium. This electron would have to come from an *inner* shell and so a huge amount of energy would be required. Even though the lattice energy for $NaCl_2$ would be more exothermic than that of NaCl, this is not enough to compensate for the large endothermic second ionisation energy. Thus ΔH_f for $NaCl_2$ would be highly endothermic and so it is not formed.

★ $MgCl_2$ or MgCl? With magnesium, the second electron comes from the outer shell and so the second ionisation energy is only slightly more endothermic than the first ionisation energy. This is compensated for by the greater lattice energy of $MgCl_2$ compared with MgCl.

Covalent bonding

★ A covalent bond is the attraction of the two nuclei for a shared pair of electrons.

★ Covalent substances may be:
- giant atomic — for example, diamond, graphite and quartz (SiO_2)
- simple molecular —for example, I_2 and many organic substances
- non-crystalline — for example, polymers, such as poly(ethene)

★ In a giant atomic solid, the particles are atoms that are held together by strong covalent bonds. Therefore, the melting temperature is very high.

'Dot-and-cross' diagrams

★ A single covalent bond consists of a pair of shared electrons, with each atom supplying one electron.

★ A dative covalent bond consists of a pair of shared electrons, with both electrons supplied by one atom

★ A double bond consists of two shared pairs of electrons, each atom supplying two electrons.

★ The result is that both atoms often have an octet of electrons. However, sometimes one atom has either only four or six electrons, or even more than eight electrons in its outer orbit.

★ Some examples of single covalent bonding are shown below.

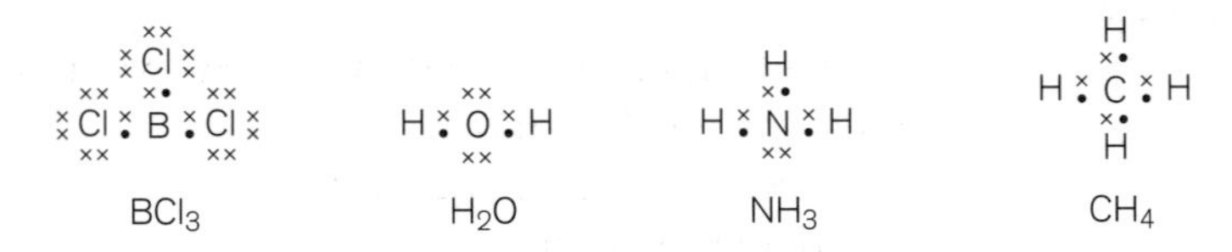

★ An example of dative covalent bonding is shown below.

NH_4^+

★ Two examples of double bonding are shown below.

$H_2C = CH_2$ $O = C = O$

Metallic bonding

★ The structure of a metal is a regular lattice of positive ions in a sea of delocalised electrons.

★ The metallic bond is the force of attraction between the positive ions and the delocalised electrons.

★ The smaller the radius of the metal ion, the stronger is the bond and the higher is the melting point.

Hint

Remember that *metals* conduct by movement of *electrons*, whereas molten or aqueous *ionic* compounds conduct by the movement of *ions*.

★ The larger the charge on the metal ion, the stronger is the bond and the higher is the melting point.

★ Metals conduct electricity because the delocalised electrons are able to move through the structure under an applied potential.

Checklist

Before attempting the questions on this topic, check that you understand:

☐ the nature of ionic, covalent and dative covalent bonds
☐ the relative sizes of positive and negative ions
☐ how to draw 'dot and cross' diagrams for ions and for covalent molecules
☐ the Born–Haber cycle
☐ polarisation of anions by cations and the effect of this on lattice energies
☐ the metallic bond and why metals conduct electricity

Testing your knowledge and understanding

For the first set of questions, cover the margin, write your answers, then check to see if you are correct.

★ State which types of covalent bond are present in a double bond, as in O=C=O.

★ State the value of the charge on the ions of the following elements:
Al, Mg, Na, N, O, F

★ **a** State which of the cations in your answer to the previous question
(i) has the smallest radius
(ii) has the greatest charge density
(iii) is the most polarising

b State which of the anions in your answer to the previous question is the most polarisable.

★ Which of Li^+, Na^+ and K^+ has the largest radius?

★ State which of the cations Li^+, Na^+ and K^+ is the most polarising.

★ State which type of particle moves when:
a a solid metal conducts electricity
b a molten ionic compound conducts electricity

Answers

σ-bond and π-bond

Al^{3+}, Mg^{2+}, Na^+, N^{3-}, O^{2-}, F^-

a (i) Al^{3+}
(ii) Al^{3+}
(iii) Al^{3+}
b N^{3-}

K^+

Li^+

a electrons
b ions

The **answers** to the **numbered questions** are on page 129

1 Draw diagrams of two overlapping p-orbitals that produce:
a a σ-bond
b a π-bond

2 What is meant by an ionic bond?

3 The ions, K^+, Ca^{2+}, S^{2-} and Cl^- are isoelectronic.
a What is meant by the term isoelectronic?
b Place these four ions in order of increasing radius.

4 Draw diagrams showing all the electrons in:
a a Li^+ ion
b an S^{2-} ion

5 A Born–Haber cycle for potassium fluoride is shown below:

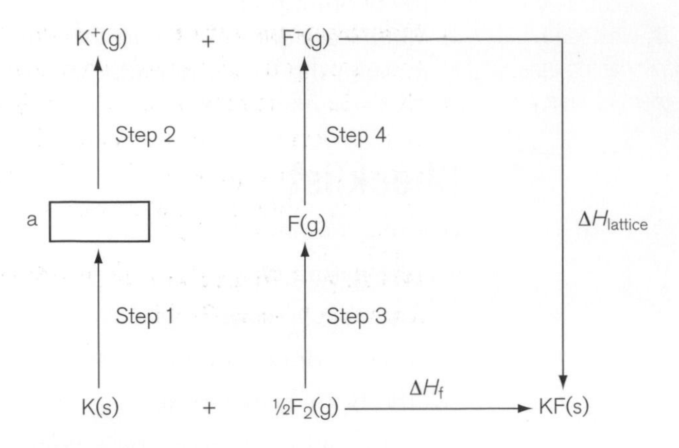

a What species is in the box labelled a?
b Identify the energy changes in steps 1, 2, 3, and 4.
c Use a data book to find the values for the energy changes in steps, 1, 2, 3 and 4 and for the value of ΔH_f of KF. Hence calculate the lattice energy of potassium fluoride.

6 Explain what is meant by a covalent bond.

7 Draw a dot-and-cross diagram (showing outer electrons only) for:
a oxygen fluoride, OF_2
b oxygen, O_2

8 Explain what is meant by a metallic bond.

9 Explain why solid sodium metal conducts electricity whereas solid sodium chloride does not.

Topic 1.5 Introductory organic chemistry

Introduction

★ An organic compound consists of a chain of one or more carbon atoms and may contain one or more functional groups (see Table 1.1). The functional group gives the compound certain chemical properties. For instance, the C=C group reacts in a similar way in all compounds. Thus, knowledge of the chemistry of ethene, $H_2C{=}CH_2$, enables you to predict the reactions of all compounds containing the C=C group.

Hint

You must learn the equations and conditions for the reactions in the specification (see pages 150–153).

Substance	Alkene	Alcohol	Aldehyde	Ketone	Acid	Ester	Nitrile	Amine
Group	>C=C<	–C–OH	H(R)C=O	R(R')C=O	–C(=O)OH	R–C(=O)O–R'	–C≡N	–C–NH₂

Table 1.1 Functional groups

Things to learn

* **Homologous series:** a series of compounds with the same functional group and the same general formula in which one member differs from the next by CH_2.
* **Structural formula:** an unambiguous formula that shows the relative positions of the atoms in the molecule. Side chains are written in brackets. The structural formula of propane is $CH_3CH_2CH_3$ and that of propan-2-ol is $CH_3CH(OH)CH_3$. It is advisable to show any double bonds, so the structural formula of ethene is $H_2C{=}CH_2$.
* **Displayed formula:** this shows all the atoms and all the bonds. The displayed formula of butan-2-ol is:

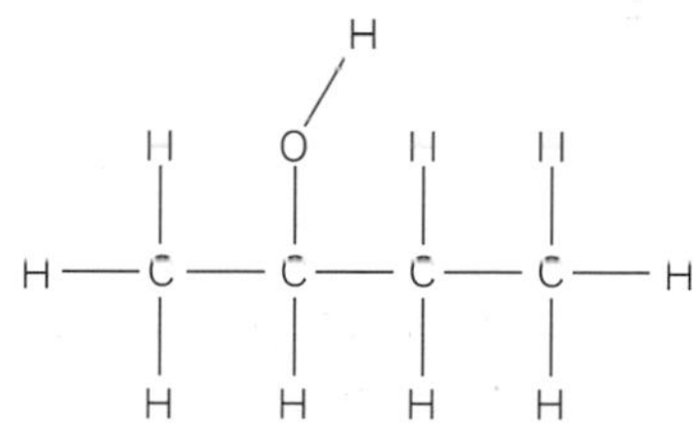

* **Skeletal formula:** this is normally used only for large molecules. The carbon skeleton is shown by zigzag lines, where each end of a line and each angle represents a carbon atom. The skeletal formula of butan-2-ol is:

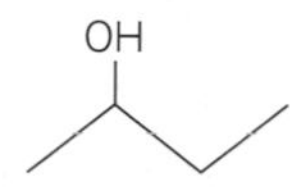

* **Homolytic fission:** when a bond breaks and one electron goes to each atom to form radicals.
* **Heterolytic fission:** when a bond breaks and the two electrons go to one atom resulting in positive and negative charged ions.
* **Substitution:** a reaction in which an atom or group of atoms in one molecule is replaced by another atom or group of atoms.
* **Addition:** a reaction in which two molecules react together to form a single product.
* **Free radical:** a species that has a single unpaired electron, e.g. Cl•
* **Electrophile:** a species that seeks out electron-rich areas and accepts a pair of electrons, forming a covalent bond.

Hint

When you write structural and displayed formulae, check that each carbon atom has four bonds, each oxygen atom has two bonds and each hydrogen atom and each halogen atom has only one bond.

Things to understand

Hazard and risk

Hazard is a property of the chemical — for example, the hazard associated with concentrated sulfuric acid is that it is corrosive. The risk is what may happen to someone using the sulfuric acid.

A **hazard** exists if the chemical is:

* corrosive
* irritant
* toxic
* absorbed through the skin
* flammable

Risks can be reduced by:

* using smaller quantities

- taking specific precautions, such as wearing gloves
- carrying out the experiment in a fume cupboard
- using a water bath or electric heater rather than a naked Bunsen flame
- using less hazardous alternative substances

Isomerism

There are three structural isomers of molecular formula C_3H_6O:
propan-1-ol $CH_3CH_2CH_2OH$
propan-2-ol $CH_3CH(OH)CH_3$
methoxyethane $CH_3OCH_2CH_3$

- **Isomers**: two or more compounds with the same molecular formula.
- **Structural isomers** may have:
  - different carbon chains (straight or branched)
  - the functional group in different places in the carbon chain
  - different functional groups
- **Geometric isomerism** occurs when there is a C=C in the molecule and each carbon has a different atom or group attached to it.
- Geometric isomers are not easily interconverted because of the restricted rotation about the π-bond.
- Naming geometric isomers:
  - If there are only two different groups out of the four around the C=C, the *cis–trans* method can be used. The *cis*-isomer has the same groups on the same side and the *trans*-isomer has them on different sides. The geometric isomers of but-2-ene are shown in the margin.
  - If there are three or four groups around the C=C, then the *E–Z* method must be used. The atoms or groups are given priorities based on atomic mass. If the atoms or groups with the highest priorities are on the same side, the isomer is called the *Z*-isomer. Thus, isomer **X** below is *Z*-3-methylpent-2-enoic acid and isomer **Y** is *E*-3-methylpent-2-enoic acid Their skeletal and structural formulae are shown below.

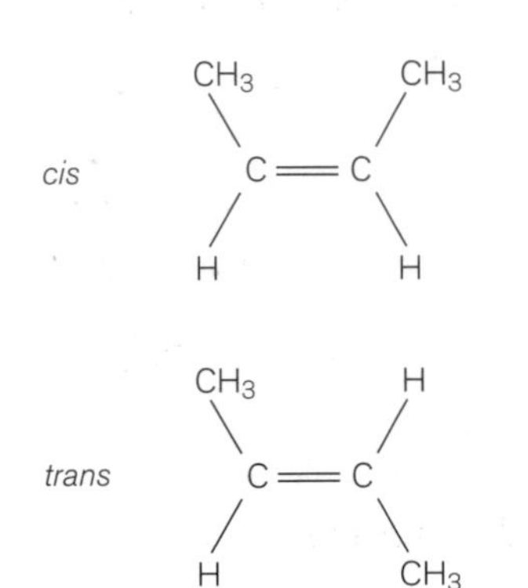

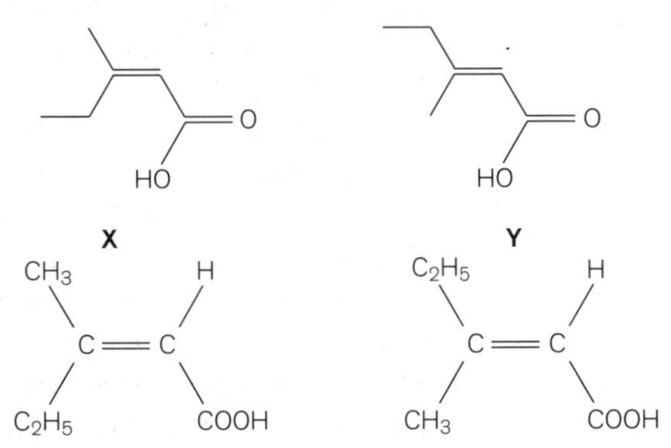

The atom or group with the larger atomic or formula mass has the higher priority. Thus C_2H_5 has a higher priority than CH_3 and COOH is higher than H.

Alkanes

- Alkanes have the general formula $C_nH_{(2n+2)}$.
- They are obtained by the fractional distillation of crude oil.

Reactions

- Alkanes burn in *excess* oxygen to form carbon dioxide and water:
  $CH_4 + 2O_2 \rightarrow CO_2 + 2H_2O$
  They are used as fuels, but as they produce carbon dioxide, their use is a cause of global warming. Alternative biofuels are being developed (page 54).
- Alkanes react with **chlorine** (or bromine) in the presence of UV light in a stepwise substitution reaction:
  $CH_4 + Cl_2 \rightarrow CH_3Cl + HCl$
  $CH_3Cl + Cl_2 \rightarrow CH_2Cl_2 + HCl$ and so on

$Cl{-}Cl \rightarrow 2Cl\bullet$

This is an example of a free radical substitution reaction.

Step 1 (initiation): The energy of UV light causes the Cl–Cl bond to break homolytically producing two chlorine free radicals.

Step 2 (propagation): The chlorine radical removes a hydrogen atom from the methane producing a hydrogen chloride molecule and a methyl radical:

$Cl\bullet + CH_4 \rightarrow HCl + CH_3\bullet$

A methyl radical then removes a chlorine atom from a chlorine molecule:

$CH_3\bullet + Cl_2 \rightarrow CH_3Cl + Cl\bullet$ and so on

Step 3 (termination):

$CH_3\bullet + CH_3\bullet \rightarrow C_2H_6$

or $CH_3\bullet + Cl\bullet \rightarrow CH_3Cl$

or $Cl\bullet + Cl\bullet \rightarrow Cl_2$

Note that in each of the propagation steps one radical produces another radical, whereas in a termination step two radicals form a molecule.

Alkenes

★ Alkenes have the general formula C_nH_{2n} and contain a C=C group, which is a σ-bond and a π-bond between the two carbon atoms.

Reactions

★ Addition of hydrogen

Equation: $CH_2{=}CH_2 + H_2 \rightarrow CH_3CH_3$

Conditions: pass gases over a heated nickel catalyst

Product: ethane

★ Electrophilic addition of bromine (or chlorine)

Equation: $CH_2{=}CH_2 + Br_2 \rightarrow CH_2BrCH_2Br$

Conditions: room temperature with the halogen dissolved in hexane

Product: 1,2-dibromoethane

Bromine water is a mixture of dissolved bromine and the products of the reaction with water.

★ Electrophilic addition of aqueous bromine (bromine water)

$Br_2 + H_2O \rightleftharpoons HOBr + HBr$

Electrophile: HOBr is the electrophile that adds on to the alkene:

$CH_2{=}CH_2 + HOBr = CH_2BrCH_2OH$

Organic product: 2-bromoethanol

The decolorisation of a red-brown solution of bromine is a test for a C=C group.

★ Electrophilic addition of hydrogen halides, e.g. HI

Equation: $CH_3CH{=}CH_2 + HI \rightarrow CH_3CHICH_3$

Conditions: mix gases at room temperature.

Product: 1-iodopropane

Hint

With the addition of HBr to unsymmetrical alkenes, such as propene, the hydrogen goes to the carbon that has more hydrogen atoms directly bonded to it. This is Markovnikoff's rule.

★ Alkenes are **oxidised** by potassium manganate(VII) solution

Equation: $CH_2{=}CH_2 + [O] + H_2O \rightarrow CH_2(OH)CH_2OH$

Conditions: room temperature, mixed with sodium hydroxide solution

Observation: a brown precipitate (of MnO_2)

Product: ethane-1,2-diol

★ Alkenes can be **polymerised.** Propene forms poly(propene) and ethene forms poly(ethene)

Equation: $nCH_2{=}CH_2 \rightarrow -(CH_2{-}CH_2)_n-$

Conditions: either a very high pressure (about 1000 atm) and a temperature of about 250°C or a catalyst of titanium(IV) chloride and triethyl aluminium at 50°C and pressure of about 10 atm.

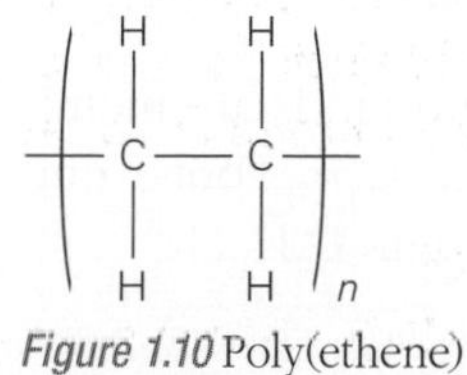

Figure 1.10 Poly(ethene)

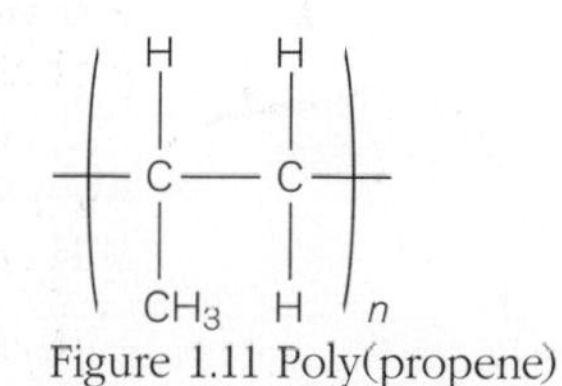

Figure 1.11 Poly(propene)

Mechanism of electrophilic addition

★ Addition of bromine to propene:

A curly arrow represents the movement of a pair of electrons.

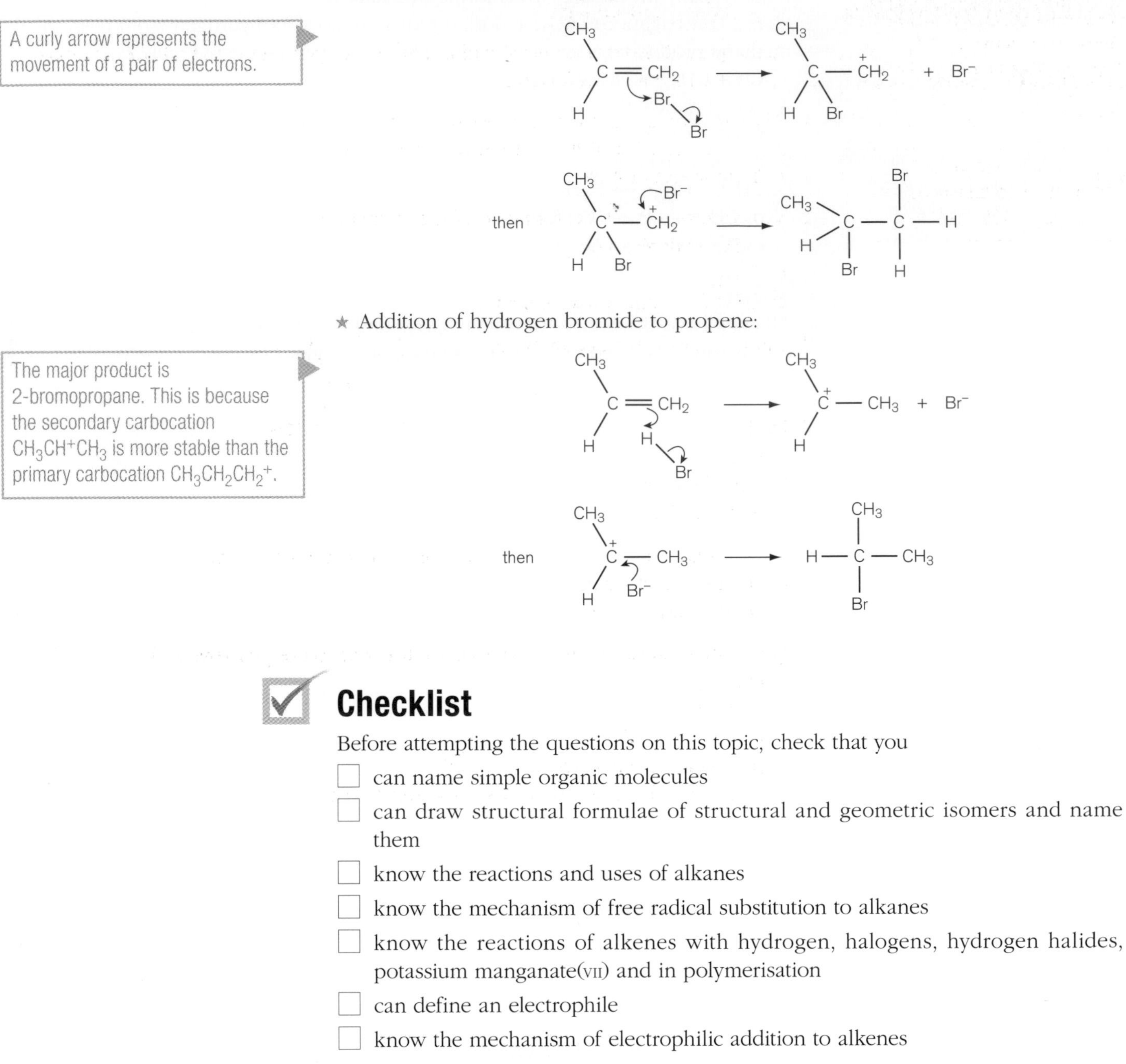

The major product is 2-bromopropane. This is because the secondary carbocation $CH_3CH^+CH_3$ is more stable than the primary carbocation $CH_3CH_2CH_2^+$.

Checklist

Before attempting the questions on this topic, check that you

- [ ] can name simple organic molecules
- [ ] can draw structural formulae of structural and geometric isomers and name them
- [ ] know the reactions and uses of alkanes
- [ ] know the mechanism of free radical substitution to alkanes
- [ ] know the reactions of alkenes with hydrogen, halogens, hydrogen halides, potassium manganate(VII) and in polymerisation
- [ ] can define an electrophile
- [ ] know the mechanism of electrophilic addition to alkenes

Testing your knowledge and understanding

For the first set of questions, cover the margin, write your answers, then check to see if you are correct.

★ Which of the following compounds are members of the same homologous series: ethene C_2H_4, cyclopropane C_3H_6, but-1-ene $CH_3CH_2CH=CH_2$, buta-1,3-diene $CH_2=CH-CH=CH_2$, cyclohexene C_6H_{10} ?

Answers

Ethene and but-1-ene

Answers
2-chloro-3-methylpentan-1-ol
Tetrachloromethane, CCl_4
C_2H_4O

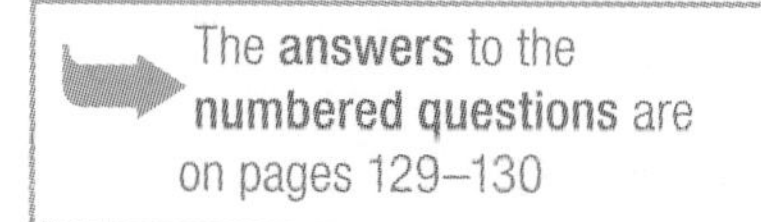
The **answers** to the **numbered questions** are on pages 129–130

★ Name $CH_3CH_2CH(CH_3)CHClCH_2OH$

★ What product is obtained if a large excess of chlorine is mixed with methane and exposed to diffused light?

★ A compound X contains 54.5% carbon, 36.4% oxygen and 9.1% hydrogen by mass. Calculate its empirical formula.

1 Write the structural formula of each of the following:
- **a** 1,1-dibromo-1,2-dichloro-2,3-difluoropropane
- **b** 1-chlorobutan-2-ol

2 Write out the structural formulae of the isomers of
- **a** C_3H_8O (alcohols only)
- **b** C_5H_{12}
- **c** C_4H_8 (no cyclic compounds)

3 Name the following compound:

CH_3 and Cl on one carbon, H and $COOH$ on the other: $(CH_3)(Cl)C{=}C(H)(COOH)$

4 Define:
- **a** free radical
- **b** homolytic fission
- **c** electrophile

5 Write equations and give conditions for the reaction between:
- **a** ethane and chlorine
- **b** ethane and oxygen

6 Write equations and give conditions for the reaction of propene with:
- **a** hydrogen
- **b** bromine*
- **c** hydrogen iodide
- **d** potassium manganate(VII) solution*

For reactions marked *, state what you would see.

Practice Unit Test 1

Section A

1 Which of the following is a balanced ionic equation?

A $Sn^{2+}(aq) + Fe^{3+}(aq) \longrightarrow Sn^{4+}(aq) + Fe^{2+}(aq)$

B $Ba^{2+}(aq) + 2Cl^-(aq) + Cu^{2+}(aq) + SO_4^{2-}(aq) \longrightarrow BaSO_4(s) + Cu^{2+}(aq) + 2Cl^-(aq)$

C $Na(s) + H^+(aq) \longrightarrow Na^+(aq) + \frac{1}{2}H_2(g)$

D $Na_2O(s) + H_2O(l) \longrightarrow 2Na^+(aq) + 2OH^-(aq) + \frac{1}{2}H_2(g)$

2 The correct definition of relative atomic mass of an element is

A the average mass of the element relative to 1/12 the mass of a carbon atom

B the average mass of the atoms of the element relative to 1/12 the mass of a carbon-12 atom

C the mass of 1 atom of the element relative to 1/12 the mass of a carbon-12 atom

D the average mass of the element relative to 1/12 the mass of a carbon-12 atom

3 The molar mass of aluminium hydroxide, $Al(OH)_3$ is

A 22 B 40 C 44 D 78

4 The amount (moles) of sodium hydroxide in $25\,cm^3$ of a $0.20\,mol\,dm^{-3}$ solution of sodium hydroxide is

A 0.0050 B 0.0500 C 5.0 D 125

5 A sample of air contained 0.33 ppm methane. The mass of air that would contain 1.0 tonne of methane is

A 3.0×10^6 tonne B 3.3×10^6 tonne C 3.3×10^{11} tonne D 3.0×10^{12} tonne

6 2-bromoethanol is prepared according to the reaction:

$C_2H_4 + Br_2 + H_2O \longrightarrow C_2H_4(OH)Br + HBr$

The atom economy in the preparation is

A 33% B 61% C 82% D 100%

7 The number 0.001230 is written to

A 3 significant figures B 4 significant figures
C 5 significant figures D 6 significant figures

8 What is the number of fluoride ions in 100 ng (nanograms) of calcium fluoride, CaF_2? (The Avogadro number is $6.02 \times 10^{23}\,mol^{-1}$)

A 7.71×10^{14} B 1.54×10^{15} C 7.71×10^{23} D 1.54×10^{24}

9 The enthalpy of atomisation of iodine is represented by

A $I_2(s) \longrightarrow 2I(g)$ B $I_2(g) \longrightarrow 2I(g)$ C $\frac{1}{2}I_2(s) \longrightarrow I(g)$ D $\frac{1}{2}I_2(g) \longrightarrow I(g)$

10 What is the enthalpy change for the following reaction?

$CH_2{=}CHCH{=}CH_2 + 2HBr \longrightarrow CH_3CHBrCHBrCH_3$

ΔH_f /kJ mol^{-1} for $CH_2{=}CHCH{=}CH_2$ is x, for HBr is y and for $CH_3CHBrCHBrCH_3$ is z

A $(x + y) - z$ B $(x + 2y) - z$ C $z - (x + y)$ D $z - (x + 2y)$

11 A solution containing 10 g of silver nitrate (molar mass $169.9\,g\,mol^{-1}$) was reacted with excess potassium carbonate solution and 8.0 g of silver carbonate (molar mass $275.8\,g\,mol^{-1}$) was produced. The equation for the reaction is:

$2AgNO_3 + K_2CO_3 \longrightarrow Ag_2CO_3 + 2KNO_3$

The percentage yield was

A 40% B 49% C 80% D 99%

12 Bromine water reacts with ethene to form

A $CHBr_2CH_3$ B CH_2BrCH_2Br C $CH_2(OH)CH_2Br$ D $CH(OH)BrCH_3$

13 Liquid bromine is a corrosive liquid with an irritating vapour. Which of the statements below is true?

A The hazard due to bromine is too great for it to be used in a school laboratory.

B The risk of using bromine is too great for it to be used in a school laboratory.

C The risk of using bromine is reduced by wearing gloves and carrying out the experiment in a fume cupboard.

D The hazard of using bromine is reduced by wearing gloves and carrying out the experiment in a fume cupboard.

14 The reaction between methane and chlorine requires UV light. This is needed

A to provide the energy to break a Cl–Cl bond and produce chlorine radicals

B to provide the energy to break a C–H bond and produce methyl and hydrogen radicals

C to enable chlorine radicals to remove a hydrogen atom from methane

D to enable hydrogen radicals to remove a chlorine atom from chlorine

15 What is the name of the following compound?

```
CH3         H
   \       /
    C === C
   /       \
HO          COOH
```

A *E*-3-hydroxybut-2-enedioc acid

B *Z*-3-hydroxybut-2-enedioc acid

C *E*-2-hydroxybut-3-enedioc acid

D *Z*-2-hydroxybut-3-enedioc acid

16 S^{2-}, Cl^-, K^+ and Ca^{2+} are isoelectronic. The largest of these ions is

A S^{2-} **B** Cl^- **C** K^+ **D** Ca^{2+}

17 Solid sodium chloride does not conduct electricity but aqueous sodium chloride does. This is because

A in solid sodium chloride the electrons are not free to move because they are fixed in the lattice between the ions but the ions are free to move in the aqueous solution

B in solid sodium chloride the ions are not free to move because they are fixed in the lattice but the ions are free to move in the aqueous solution

C in solid sodium chloride the electrons are not free to move because they are fixed in the lattice between the ions but in aqueous sodium chloride the electrons are free to move

D in aqueous sodium chloride the sodium chloride has reacted with the water to produce sodium hydroxide and hydrochloric acid, both of which conduct

18 $MgCl_3$ does not exist because

A Mg^{3+} ions do not have the stability of a noble gas

B magnesium can only lose two electrons

C the energy needed to remove the third electron is not compensated for by that released when the lattice is formed from the gaseous ions

D the energy needed to remove the third electron is not compensated for by that required to form the lattice from the gaseous ions

19 The successive ionisation energies ($kJ\,mol^{-1}$) of an element X are shown in the table below.

1st	2nd	3rd	4th	5th	6th	7th	8th
1060	1900	2920	4960	6280	21 200	25 900	30 500

This is evidence for X being in which group of the periodic table?

A 4 **B** 5 **C** 6 **D** 7

20 Bromine consists of two isotopes, ^{79}Br and ^{81}Br. Its relative atomic mass is 80. Its mass spectrum will have peaks at *m/e* values of

A 79 and 81 only **B** 79, 80 and 81 **C** 79, 81, 158 and 162 **D** 79, 81, 158, 160 and 162

Section A total: 20 marks

Section B

21 a Define the term 'standard enthalpy of formation'. **(3 marks)**

b Use the data and the Hess's law cycle below to calculate the enthalpy of formation of ethane, $C_2H_6(g)$.

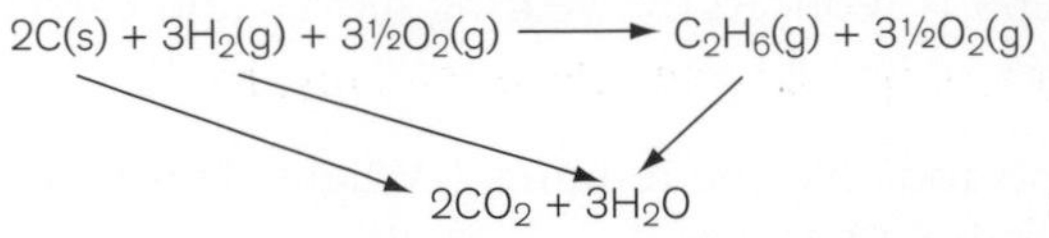

$$2C(s) + 3H_2(g) + 3\tfrac{1}{2}O_2(g) \longrightarrow C_2H_6(g) + 3\tfrac{1}{2}O_2(g)$$

$$2CO_2 + 3H_2O$$

Enthalpy of combustion data: ethane = –1560 kJ mol^{-1}; carbon = –394 kJ mol^{-1}; hydrogen = –286 kJ mol^{-1} **(3 marks)**

c The enthalpy of combustion of ethane can be estimated by burning a known volume of ethane beneath a copper beaker containing water. The results of such an experiment are shown below:

volume of ethane burnt: 135 cm^3
mass of water heated up: 100 g
temperature rise of water: 19°C
specific heat capacity of water: 4.18 J K^{-1} g^{-1}
molar volume of gas: 24 000 cm^3 mol^{-1}

Calculate the enthalpy of combustion of ethane. Include a sign and units in your answer. **(5 marks)**

d Ethene and hydrogen react to form ethane according to the equation

$$CH_2{=}CH_2(g) + H_2(g) \rightarrow C_2H_6(g)$$

Given the bond enthalpies below, calculate the enthalpy change of this reaction.

C=C 612 kJ mol^{-1}; C–C 348 kJ mol^{-1}; C–H 412 kJ mol^{-1} H–H 436 kJ mol^{-1} **(4 marks)**

Total: 15 marks

22 a How could you demonstrate that the purple potassium manganate(VII) is ionic? **(3 marks)**

b Draw a dot-and-cross diagram, showing the outer electrons only, of

(i) a calcium ion **(1 mark)**

(ii) a chloride ion **(1 mark)**

c The Born–Haber cycle for the formation of calcium chloride is shown below.

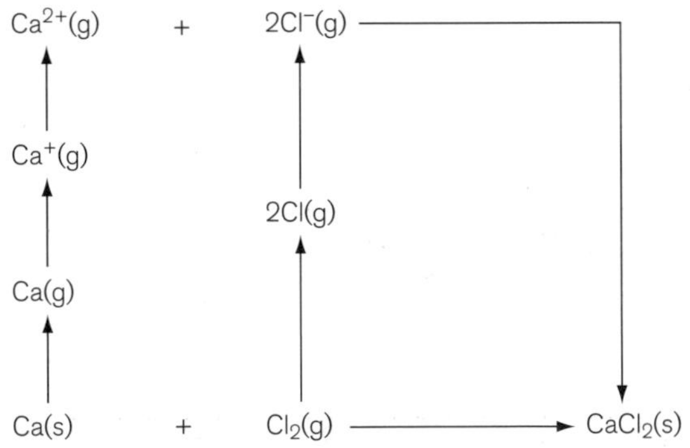

$Ca^{2+}(g)$ + $2Cl^-(g)$

$Ca^+(g)$

$2Cl(g)$

$Ca(g)$

$Ca(s)$ + $Cl_2(g)$ ⟶ $CaCl_2(s)$

Data:

ΔH_a(calcium) +193 kJ mol^{-1}
ΔH_a(chlorine) +121 kJ mol^{-1}
1st IE (calcium) +590 kJ mol^{-1}
2nd IE (calcium) +1150 kJ mol^{-1}
EA (chlorine) –364 kJ mol^{-1}
$\Delta H_{formation}$ $CaCl_2(s)$ –795 kJ mol^{-1}

(i) Calculate the lattice energy of calcium chloride. Include a sign and unit in your answer. **(4 marks)**

(ii) Explain why $CaCl_3$ does not exist. **(3 marks)**

(iii) The theoretical lattice energy for calcium chloride is –2200 kJ mol^{-1}. Explain why this value differs from that calculated from the experimental value obtained from the Born–Haber cycle. **(2 marks)**

(iv) Would you expect the difference to be greater or smaller with calcium bromide? Justify your answer. **(2 marks)**

d Define a covalent bond. **(1 mark)**

e Draw dot-and-cross diagrams, showing outer electrons only, for:

(i) nitrogen, N_2 **(1 mark)**

(ii) silicon tetrachloride, $SiCl_4$ **(1 mark)**

Total: 19 marks

23 a (i) What is meant by the term 'homologous series'? **(2 marks)**

(ii) Methane and octane are both members of the homologous series of alkanes and are both used as fuels.

Calculate the volume of oxygen needed to burn 10 g of octane according to the equation:

$C_8H_{18}(l) + 12\frac{1}{2}O_2(g) \rightarrow 8CO_2(g) + 9H_2O(l)$

(The molar volume of a gas under these conditions is 24 $dm^3\ mol^{-1}$) **(4 marks)**

b (i) Write the equation for the reaction between methane and chlorine. **(1 mark)**

(ii) Draw the mechanism for this reaction, showing the initiation and propagation steps and one termination process. **(4 marks)**

c (i) Ethene and propene are members of the homologous series of alkenes. Describe the test for alkenes. **(2 marks)**

(ii) Write the equation for the reaction of propene with bromine dissolved in hexane. **(1 mark)**

(iii) Draw the mechanism for the reaction between propene and hydrogen bromide. **(3 marks)**

(iv) Explain why the major product is 2-bromopropane and not 1-bromopropane. **(2 marks)**

Total: 19 marks

24 A sketch of the first ionisation energies of elements 10 to 19 (neon to potassium) is shown below.

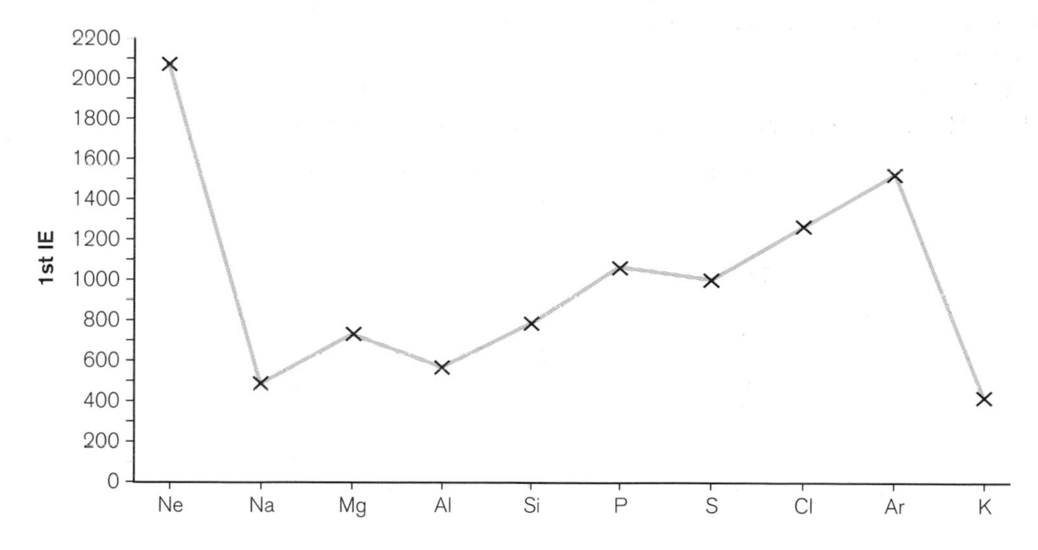

Explain why:

a there is a general upward trend from sodium to argon **(3 marks)**

b there is a slight decrease in ionisation energy between magnesium and aluminium **(1 mark)**

c there is a big drop after each noble gas **(1 mark)**

d the first ionisation energy of potassium is less than that of sodium. **(2 marks)**

Total: 7 marks

Section B total: 60 marks

Paper total: 80 marks

UNIT 2

Application of core principles of chemistry

Topic 2.1 Shapes of molecules and ions, polarity and intermolecular forces

Introduction

* The shapes of molecules and ions are determined by the **valence shell electron pair repulsion** (VSEPR) theory.
* This states that electron pairs around an atom, in bonds and in lone pairs, repel each other to maximise the distance between them, thus minimising repulsion.
* Intermolecular forces are the forces *between* covalent molecules.

Things to learn

* The **electronegativity** of an element is a measure of the attraction an atom has on a pair of electrons in a covalent bond.
* A **hydrogen bond** is the attraction between a δ+ H in one molecule and the **lone pair** of electrons in a δ– F, O or N in another molecule.
* The number of **lone pairs** around an atom can be worked out using the formula:

  number of lone pairs = ½ (group number of the element – the number of covalent bonds)

  Examples:
  - O in H_2O: group = 6, number of bonds = 2, lone pairs = ½(6 – 2) = 2
  - S in SO_2: group = 6, bonds = 4 (2 double), lone pairs = ½(6 – 4) = 1
* **Bond length** is the distance between the two nuclei.

Things to understand

Shapes of molecules and ions

* The bond pairs and lone pairs repel so as to take up a position where their repulsion is a minimum.

Bond pairs	Lone pairs	Example	Shape	Bond angle/°	Diagram
2	0	$BeCl_2$	Linear	180	180°
3	0	BCl_3	Triangular planar	120	120°

Bond pairs	Lone pairs	Example	Shape	Bond angle/°	Diagram
4	0	CH_4, NH_4^+ and the $-CH_2OH$ group in alcohols	Tetrahedral	109.5	109.5°
4	1	NH_3	Triangular pyramidal	107	107°
4	2	H_2O	Bent (V-shaped)	104.5	< 105°
5	0	PCl_5	Trigonal bipyramidal	120 and 90	120° 90°
6	0	SF_6	Octahedral	90 and 180	90 and 180°

Hint

Remember that a double bond consists of one σ-bond one π-bond.

★ Molecules with double bonds — count the number of σ-bonds and the number of lone pairs then deduce the shape and bond angle as in the table below.

Example	σ-bonds	Lone pairs	Shape	Bond angle
O=C=O	2	0	Linear	180
O=S=O	2	1	Bent	120
$H_2C{=}CH_2$	3	1	Triangular planar (both carbons)	120

Effect of lone pairs on bond angle

★ Lone pair/lone pair repulsion is greater than lone pair/bond pair repulsion, which is greater than bond pair/bond pair repulsion. This extra repulsion causes the bond angle to decrease by about 2.5° per lone pair.

★ NH_3 has three bond pairs and one lone pair. The electron pairs are arranged in a tetrahedron, so the shape is pyramidal with the bond angle 2.5° less than 109.5°.

The lone pairs contribute to the distribution of electrons in space, but the shape is determined by the arrangement of the atoms.

★ H_2O has two bond pairs and two lone pairs. The electron pairs are arranged in a tetrahedron, so the molecule is V-shaped with the bond angle 5° less than 109.5°.

Shapes of ions

Negative ions have gained one electron for each negative charge, which are on oxygen atoms in SO_4^{2-}, CO_3^{2-} and NO_3^-:

★ SO_4^{2-}: the sulfur has four σ-bonds (and two π-bonds) and no lone pairs, so the ion is tetrahedral.

★ CO_3^{2-}: the carbon has three σ-bonds (and one π-bond) and no lone pairs and so the ion is triangular planar.

- NO_3^-: there is one single covalent, one double covalent and one dative covalent bond (3 σ-bonds) and no lone pairs. It is triangular planar.
- Positive ions have lost one electron for each positive charge:
- NH_4^+: nitrogen has four electrons and forms four σ-bonds and no lone pairs. The ion is tetrahedral.

Shapes of forms of carbon

- **Graphite**: each carbon is σ-bonded to three other carbon atoms in a plane, with the fourth ($2p_z$) electron delocalised above and below the plane.
- **Diamond**: each carbon is σ-bonded to four other carbon atoms in a giant tetrahedral structure.
- **Fullerenes**: each carbon is σ-bonded to three other carbon atoms, often in a 60 atom ball. The fourth electron is delocalised through the structure.
- **Nanotubes**: these are the result of wrapping a one-atom-thick layer of graphite into a seamless cylinder with an internal diameter as little as a few nanometres. One end is capped by a hemisphere of a fullerene.
- Graphite, fullerenes and nanotubes all conduct electricity using delocalised electrons.

Polarity

Polar covalent bonds

If the two elements in a covalent bond have different electronegativities, the bond is polar. The more electronegative atom is δ– and the less electronegative atom is δ+, so, for example, the H in HBr is δ+ and the Br is δ–.

Polar molecules

If a covalent molecule has polar bonds, the molecule as a whole is polar if the dipoles do not cancel. The C=O bonds in CO_2 and the C–H bonds in CH_4 are polar, but as the molecules are symmetrical, the dipoles cancel and the molecules are non-polar.

The H–O bonds in water are polar, but as the molecule is V-shaped the dipoles do not cancel. Therefore, the molecule is polar and has a dipole moment.

Intermolecular forces

These occur *between* different covalent molecules and are much weaker than covalent bonds. There are three types:

- **Hydrogen bonds** are the strongest type of intermolecular force. They occur between a δ+ hydrogen atom in one molecule and the lone pair on a small δ– atom on another molecule. Hydrogen bonding occurs, for example, in HF, H_2O, C_2H_5OH and NH_3. The bond angle around the hydrogen is 180°.

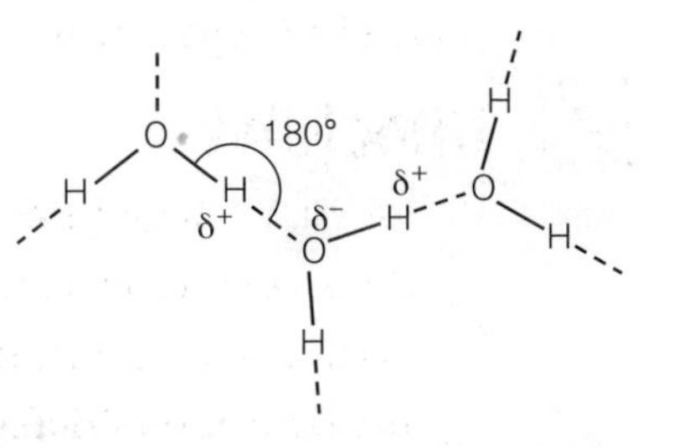

Figure 2.1 Hydrogen bonding in water

London forces are sometimes called dispersion forces.

- **Induced dipole (London) forces** occur between all covalent molecules. Their strength depends mainly on the number of electrons in the molecules. They are caused by oscillating electric fields in the molecules.
  Permanent dipoles occur between the δ+ part of one molecule and the δ− part of another. They are weaker than London forces between molecules with a similar number of electrons.

Boiling and melting temperatures

Branched-chain alkanes have a lower boiling point than their straight-chain isomers due to less efficient packing. Likewise, alkenes have lower boiling points because they pack less well than alkanes.

- These depend on the strength of the intermolecular forces. Thus propane (26 electrons) has a lower boiling point than ethanol (also 26 electrons), which is hydrogen-bonded.
- The boiling points of the homologous series of alkanes increase as the number of electrons per molecule increases.
- The boiling points of the hydrogen halides HCl to HI increase as the strength of their London forces increase, despite the decrease in permanent dipole forces.
- Hydrogen fluoride, HF, is hydrogen-bonded and so has a higher boiling temperature than the other hydrogen halides.

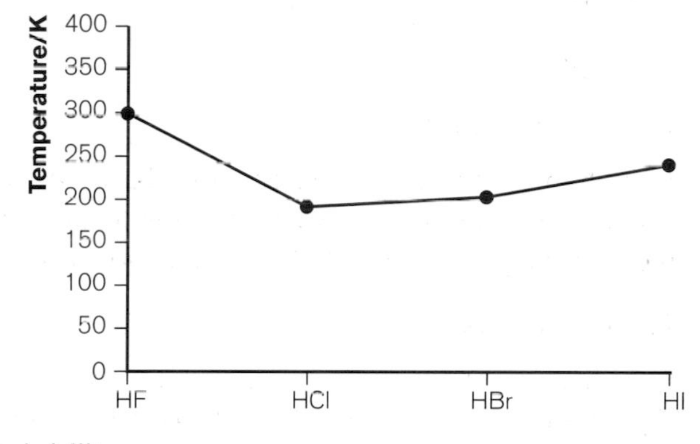

Figure 2.2 Boiling temperatures of group 7 hydrogen halides

Solubility

Ionic compounds

Many ionic compounds are water-soluble as energy is released due to the attraction between the positive ions and the δ− oxygen atoms in water and between the negative ions and the δ+ hydrogen atoms in water.

Hydrogen-bonded covalent molecules

Ethanol can form hydrogen bonds with water molecules and is, therefore, soluble in water.

Covalent molecules without hydrogen bonding

Even if polar, these do not dissolve in water. They dissolve in other non-hydrogen-bonded solvents, such as hexane, C_6H_{14}, and ethoxyethane, $C_2H_5OC_2H_5$.

Checklist

Before answering questions on this topic, check that you understand:

- [ ] the shapes of molecules and ions
- [ ] bond angles in molecules with and without lone pairs of electrons
- [ ] the structure of different forms of carbon
- [ ] that polar bonds may not give rise to polar molecules

- [ ] intermolecular forces, such as hydrogen bonding, induced dipole forces and permanent dipole forces
- [ ] the trends in boiling temperature caused by intermolecular forces
- [ ] solubility of ionic compounds, hydrogen-bonded covalent and non-hydrogen-bonded covalent compounds

Testing your knowledge and understanding

For the first set of questions, cover the margin, write your answers, then check to see if you are correct.

★ State the bond angle in
a water **b** ammonia

★ What is the bond angle around the hydrogen-bonded hydrogen atom in HF?

★ **a** Give the most electronegative element in each of the following groups:
(i) C, O, N, F (ii) Cl, Br, I
b Give the least electronegative element in each of the following groups:
(i) Be, Mg, Ca (ii) Na, Mg, Al

★ In each of **a** to **d** name the molecule whose bonds are the most polar bonds.
a HF, HCl, HBr **b** HF, H_2O, NH_3 **c** H_2O, H_2S, H_2Se
d $AlCl_3$, $SiCl_4$, PCl_5

★ Of the molecules in the table, list those that are polar

H_2	HCl	H_2O	NH_3	CH_4
CO_2	ICl	H_2S	BH_3	CF_4

★ Which is the strongest intermolecular force between HCl molecules?

★ Which of the compounds in the table have hydrogen bonds between their molecules?

H_2O	NH_3	HF	HCl	CO_2	Nylon
CH_3OH	CH_3NH_2	CH_3F	$CHCl_3$	$CH_2{=}CH_2$	Poly(ethene)

★ Arrange the noble gases Ar, He, Kr, Ne and Xe in order of *decreasing* boiling temperature.

★ Which of the compounds HF, HCl, HBr, or HI has the *lowest* boiling temperature?

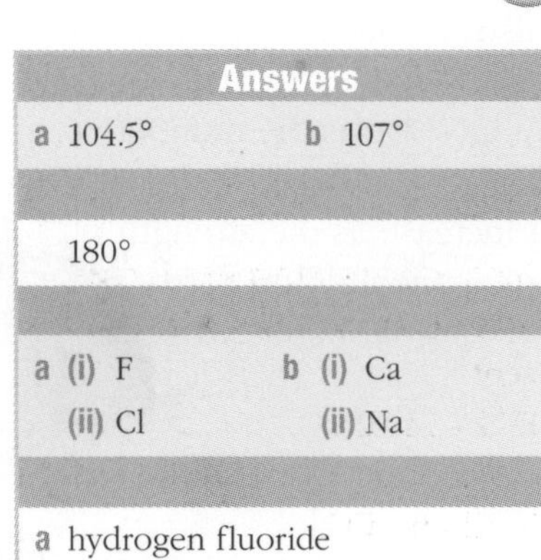

Answers

a 104.5° b 107°

180°

a (i) F b (i) Ca
(ii) Cl (ii) Na

a hydrogen fluoride
b hydrogen fluoride
c water
d aluminium chloride

HCl, H_2O, NH_3, ICl, H_2S

Induced dipole (London/dispersion) forces

H_2O, NH_3, HF, nylon, CH_3OH, CH_3NH_2

Xe > Kr > Ar > Ne > He

HCl

The **answers** to the **numbered questions** are on page 132

1 In order to work out the shape of a molecule or ion, you must first evaluate the number of σ–bonded pairs of electrons and the number of lone (unbonded) pairs of electrons around the central atom.
Construct a table similar to that shown below and use it to deduce the shapes of the following species:
SiH_4, BF_3, $HC{\equiv}CH$, PCl_3, H_2S, NH_4^+, PCl_6^-

Molecule/ion	Number of σ-bond pairs	Number of lone pairs	Total number of electron pairs	Shape
NH_3	3	1	4	Pyramidal

2 Explain why CCl_4 is non-polar, whereas $CHCl_3$ is polar.

3 Explain how a hydrogen bond forms between HF molecules.

4 Three types of Intermolecular force are:
- hydrogen bond
- dispersion (induced dipole–induced dipole)
- permanent dipole–permanent dipole

For each of the following substances state *all* the types of intermolecular force present:

a HF b I_2 c HBr d PH_3 e Ar

5 Explain why:
a ethane has a lower boiling temperature than propane
b propene has a lower boiling temperature than propane
c methyl propane has a lower boiling temperature than butane

6 For each of the following pairs of substances, state which has the stronger forces between particles and hence has the higher boiling temperature. In each case, give an explanation in terms of the types of force present.
a NH_3 and PH_3
b HCl and HBr
c CH_3COCH_3 (propanone) and C_4H_{10} (butane)
d P_4 and S_8
e NaCl and CCl_4

7 Explain why CH_3Cl is insoluble in water whereas CH_3OH is soluble, even though both are polar molecules.

Topic 2.2

Redox

Introduction

★ Redox reactions involve electron transfer.
★ Remember **oil rig** (**o**xidation **i**s **l**oss, **r**eduction **i**s **g**ain of electrons).

Things to learn

★ **Oxidation** occurs when a substance loses one or more electrons. There is an increase in the oxidation number of the element involved.
★ **An oxidising agent** is a substance that oxidises another substance and so is itself reduced. The half equation involving an oxidising agent has electrons on the left-hand side, i.e. it *takes* electrons from the substance being oxidised.
★ **Reduction** occurs when a substance gains one or more electrons. There is a decrease in the oxidation number of the element involved.
★ **A reducing agent** is a substance that reduces another substance and so is itself oxidised. The half-equation involving a reducing agent has electrons on the right-hand side, i.e. it *gives* electrons to the substance being reduced.

Things to understand

Oxidation number

★ The oxidation number is the charge on an atom of the element in a compound calculated assuming that all the atoms in the compound are simple monatomic ions. The more electronegative element is given an oxidation number of –1 per bond.

- There are some rules used for calculating oxidation numbers. They should be applied in the following order:

  Rule 1 The oxidation number of an uncombined element is zero.

  Rule 2 A simple monatomic ion has an oxidation number equal to its charge.

  Rule 3 The oxidation number of a group 1 metal is always +1; that of a group 2 metal is +2.

  Rule 4 Fluorine always has an oxidation number of –1, hydrogen of +1 (except in metallic hydrides), and oxygen of –2 (except in OF_2 and peroxides).

  Rule 5 The sum of the oxidation numbers in a molecule is 0; those in a polyatomic ion (such as SO_4^{2-}) add up to the charge on the ion.

In an overall equation, the total increase in oxidation number of one element (increase × number of atoms of that element) must equal the total decrease in oxidation number of another element.

- When an element is oxidised, its oxidation number increases.

Worked example

Calculate the oxidation number of chlorine in Cl_2, Cl^-, $MgCl_2$ and ClO_3^-.

Answer

In Cl_2: 0 (uncombined element: rule 1)
In Cl^-: –1 (monatomic ion: rule 2)
In $MgCl_2$: –1 (Mg is 2+. So: (+2) + 2Cl = 0: rule 3 and rule 5)
In ClO_3^-: +5 (Cl + 3 × (–2) = –1: rules 4 and 5)

Half-equations

These are written either:

- as reduction with electrons on the left side of the half-equation, for example:

  $Cl_2(g) + 2e^- \rightleftharpoons 2\ Cl^-(aq)$

  Here, chlorine is being reduced and is, therefore, acting as an oxidising agent.

  or
- as oxidation with electrons on the right side of the half-equation, for example:

  $Fe^{2+}(aq) \rightleftharpoons Fe^{3+}(aq) + e^-$

  Here, the iron(II) ion is being oxidised and so is acting as a reducing agent.
- Many oxidising agents only work in acid solution. Their half-equations have H^+ ions on the left-hand side and H_2O on the right. This is likely with oxidising agents containing oxygen, for example MnO_4^-:

  $MnO_4^-(aq) + 8H^+(aq) + 5e^- \rightleftharpoons Mn^{2+}(aq) + 4H_2O(l)$
- If a redox system is in alkaline solution, OH^- may need to be on one side and H_2O on the other, for example:

  $Cr^{3+}(aq) + 8OH^-(aq) \rightleftharpoons CrO_4^{2-}(aq) + 4H_2O(l) + 3e^-$

Hint

Remember that an oxidising agent gets reduced (loses electrons), and a reducing agent gets oxidised. Half-equations must balance for atoms *and* for charge.

Overall redox equations

- Overall redox equations are obtained by adding half-equations together.
- One half-equation must be written as a reduction (electrons on the left) and the other as an oxidation (electrons on the right)
- When they are added together, the number of electrons must cancel out. To achieve this it may be necessary to multiply one or both half-equations by an integer. For example, for the overall equation for the oxidation of Fe^{2+} ions by MnO_4^- ions:

  add $MnO_4^-(aq) + 8H^+(aq) + 5e^- \rightleftharpoons Mn^{2+}(aq) + 4H_2O(l)$

  to $5 \times Fe^{2+}(aq) \rightleftharpoons Fe^{3+}(aq) + e^-$

  $MnO_4^-(aq) + 8H^+(aq) + 5Fe^{2+}(aq) \rightleftharpoons Mn^{2+}(aq) + 4H_2O(l) + 5Fe^{3+}(aq)$

Hint

If the question asks for the overall equation for the reaction of A with B, make sure that *both* A and B appear on the left-hand side of the final overall equation.

Disproportionation

In a disproportionation reaction an element in a *single* species is simultaneously oxidised and reduced. For this to happen, the element must be able to have at least three different oxidation numbers.

The reaction between bromine and water is an example.

$$\underset{0}{Br_2} + H_2O \rightleftharpoons \underset{+1}{HOBr} + \underset{-1}{HBr}$$

Bromine, oxidation number 0, is oxidised to HOBr and reduced to HBr.

Checklist

Before attempting the questions on this topic, check that you can:

- ☐ define oxidation, reduction and disproportionation
- ☐ define oxidising agent and reducing agent
- ☐ calculate oxidation numbers in neutral molecules and in ions
- ☐ write ionic half-equations
- ☐ combine half-equations and so deduce an overall redox equation

Testing your knowledge and understanding

For the first set of questions, cover the margin, write your answers, then check to see if you are correct.

★ In the following equations, state which substance, if any, has been oxidised:

a $2Ce^{4+}(aq) + 2I^-(aq) \rightarrow 2Ce^{3+}(aq) + I_2(aq)$
b $H^+(aq) + OH^-(aq) \rightarrow H_2O(l)$
c $Zn(s) + 2H^+(aq) \rightarrow Zn^{2+}(aq) + H_2(g)$
d $2Fe^{2+}(aq) + 2Hg^{2+}(aq) \rightarrow 2Fe^{3+}(aq) + Hg_2^{2+}(aq)$

★ Calculate the oxidation number of the elements in bold in the following:

a $\mathbf{S}O_2$, $H_2\mathbf{S}$,
b $\mathbf{Cr}O_4^{2-}$, $\mathbf{Cr}_2O_7^{2-}$,
c $H_2\mathbf{O}_2$, $H_2\mathbf{S}O_4$, $\mathbf{Fe}_3O_4$

Answers

a iodide ions
b none
c zinc
d iron(II) ions

a +4 −2
b +6 +6
c −1 +6 +8/3 (two at +3 and one at +2)

The **answers** to the **numbered questions** are on page 132

1 a Construct ionic half-equations for the following reactions:
(i) Sn^{2+} ions being oxidised to Sn^{4+} ions in aqueous solution
(ii) Fe^{3+} ions being reduced to Fe^{2+} ions in aqueous solution
b Write a balanced equation for the reaction between Fe^{3+} and Sn^{2+} ions in aqueous solution.

2 a Construct ionic half-equations for the following reactions:
(i) hydrogen peroxide being reduced to water in acid solution
(ii) sulfur being reduced to hydrogen sulfide in acid solution
b Write the balanced equation for the reaction between hydrogen peroxide and hydrogen sulfide in acid solution.

3 a Construct ionic half-equations for
(i) $PbO_2(s)$ being reduced to $PbSO_4(s)$ in the presence of $H_2SO_4(aq)$
(ii) $PbSO_4(s)$ being reduced to $Pb(s)$ in the presence of water

b Write the balanced equation for the reaction between PbO_2 and lead in the presence of dilute sulfuric acid.

c Explain whether or not this is a disproportionation reaction.

Topic **2.3**

The periodic table: group 2

Introduction

★ The properties of elements and their compounds change steadily down the group.

★ This means that the answer to a question as to which is the most/least reactive, easiest/hardest to decompose, most/least soluble etc. will be an element, or a compound of that element, either at the top or at the bottom of the group.

★ Down the group the elements become increasingly metallic in character. Thus:

- their oxides become stronger bases
- they form positive ions more readily
- they form covalent bonds less readily

Things to learn

Physical properties of the elements

★ All are solid metals.

★ All conduct electricity.

★ Melting temperature and hardness *decrease* down the group (except for magnesium, which has a lower melting temperature than calcium). Melting temperature is higher than that of the group 1 element in the same period.

Reactions of the elements with oxygen

The vigour of the reaction increases down the group.

★ Apart from barium, which burns in excess oxygen to form the peroxide BaO_2, group 2 metals burn to form ionic oxides of formula MO, for example:

$$2Ca + O_2 \longrightarrow 2CaO$$

Reaction of the elements with chlorine

★ All react vigorously to produce ionic chlorides of formula MCl_2, except $BeCl_2$ is covalent when anhydrous.

Be^{2+} is so small that it forms the $[Be(H_2O)_4]^{2+}$ ion.

★ All group 2 chlorides are soluble in water producing hydrated ions of formula $[M(H_2O)_6]^{2+}$:

$$MCl_2(s) + aq \longrightarrow [M(H_2O)_6]^{2+}(aq) + 2Cl^-(aq)$$

where M = Mg, Ca, Sr or Ba

★ Beryllium chloride gives an acidic solution because of hydrolysis:

$$[Be(H_2O)_4]^{2+} + H_2O \rightleftharpoons [Be(H_2O)_3(OH)]^+ + H_3O^+$$

Reaction of the elements with water

★ Beryllium does not react.

★ Magnesium burns in steam to produce an *oxide* and hydrogen:

$$Mg + H_2O \longrightarrow MgO + H_2$$

The rate of reaction increases down the group.

★ The other group 2 metals react rapidly with cold water to form an alkaline suspension of metal hydroxide and bubbles of hydrogen gas, for example:

$$Ba + 2H_2O \longrightarrow Ba(OH)_2 + H_2$$

Reactions of group 2 oxides with water and dilute acids

- ★ BeO is amphoteric and does not react with water.
- ★ MgO is basic and reacts slowly with water to form a hydroxide.
- ★ The others react rapidly and exothermically to form an alkaline suspension of the hydroxide that has a pH of about 13, for example:

  $CaO + H_2O \longrightarrow Ca(OH)_2$
- ★ All group 2 oxides react with dilute acids to form a salt and water, for example:

  $CaO + 2HCl \longrightarrow CaCl_2 + H_2O$

Reactions of group 2 hydroxides with dilute acids

- ★ All react to form a salt and water, for example:

  $Ca(OH)_2 + 2HCl \longrightarrow CaCl_2 + 2H_2O$

Things to understand

Ionisation energies

The value of the first ionisation energy for the group 2 elements *decreases* down the group. This is because as the atoms get larger, the outer electrons are further from the nucleus and so are held less firmly. The increase in nuclear charge is compensated for by an increase in shielding by the greater number of inner electrons.

Solubilities of group 2 sulfates and hydroxides

- ★ Sulfates: solubility *decreases* down the group. $BeSO_4$ and $MgSO_4$ are soluble; $CaSO_4$ is slightly soluble, $SrSO_4$ and $BaSO_4$ are insoluble.
- ★ Hydroxides: solubility *increases* down the group. $Be(OH)_2$ and $Mg(OH)_2$ are insoluble, $Ca(OH)_2$ and $Sr(OH)_2$ are slightly soluble, and $Ba(OH)_2$ is fairly soluble.
- ★ Addition of aqueous sodium hydroxide to solutions of group 2 salts produces a white precipitate of metal hydroxide, for example:

  $M^{2+}(aq) + 2OH^-(aq) \longrightarrow M(OH)_2(s)$

  Barium produces only a faint precipitate.

Thermal stability of group 1 and group 2 nitrates and carbonates

- ★ This *increases* down both groups. This is because the ionic radius of the cation increases causing it to be less polarising.
- ★ The stability of a group 2 nitrate or carbonate is *less* than that of the equivalent group 1 compound. This is because the group 2 cations are 2+ whereas the group 1 cations are 1+. In addition, a group 2 cation has a smaller radius than the group 1 cation in the same period. Both these factors result in the group 2 cation being more polarising which makes decomposition easier.
- ★ Group 2 nitrates all decompose to give the metal oxide, brown fumes of nitrogen dioxide and oxygen, for example:

  $2Ca(NO_3)_2 \longrightarrow 2CaO + 4NO_2 + O_2$
- ★ Apart from lithium nitrate, group 1 nitrates decompose to give a metal nitrite and oxygen, for example:

  $2NaNO_3 \longrightarrow 2NaNO_2 + O_2$
- ★ The Li^+ ion has such a small radius that it is extremely polarising and lithium nitrate decomposes in the same way as group 2 nitrates:

  $4LiNO_3 \longrightarrow 2Li_2O + 4NO_2 + O_2$

- Apart from barium carbonate, which is stable to heat, group 2 carbonates decompose to give a metal oxide and carbon dioxide, for example:
  $CaCO_3 \rightarrow CaO + CO_2$
- Group 1 carbonates are stable to heat, except for lithium carbonate, which decomposes thus:
  $Li_2CO_3 \rightarrow Li_2O + CO_2$

Flame colours of group 1 and group 2 compounds

- Group 1: lithium carmine-red
  sodium yellow
  potassium lilac
- Group 2: calcium brick red
  strontium crimson-red
  barium apple green

These colours arise as follows:

(1) Heat causes the compound to vaporise and promotes an electron to a higher orbital.

(2) The electron falls back to its normal shell and as it does so, energy in the form of visible light is emitted. The light that is emitted is of a characteristic frequency, and hence colour, dependent on the energy level difference between the two shells (see Figure 2.3).

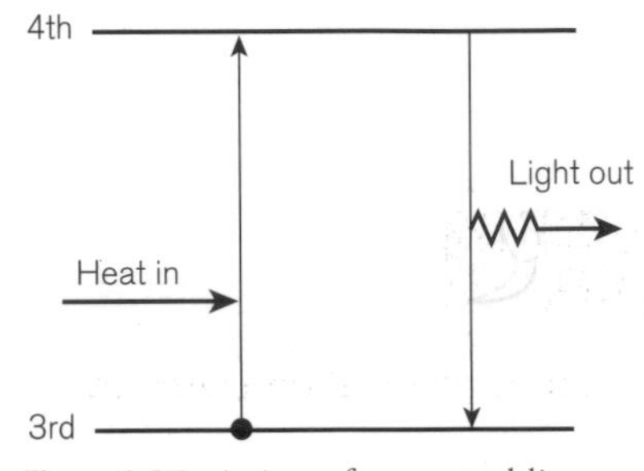

Figure 2.3 Emission of a spectral line

Titrations

Group 1 hydroxides and the hydroxides of calcium, strontium and barium are soluble in water and are, therefore, alkalis. Their concentrations can be found by titration.

- A known volume of an acid is pipetted into a conical flask and a few drops of indicator added.
- A solution of the alkali is added steadily from a burette, with mixing, until the indicator changes to the neutral colour.
- The titration is repeated until two consistent titres are obtained.

'Consistent' means that the titres must be within $\pm$ 0.2 cm^3 of each other.

Titration calculations are carried out using the following route:

$$\text{concentration and volume of A} \xrightarrow{\text{step 1}} \text{moles A} \xrightarrow{\text{step 2}} \text{moles B} \xrightarrow{\text{step 3}} \text{answer about B}$$

For steps 1 and 3, use the relationship:

amount (in moles) = $M \times V/1000$

For step 2, use the stoichiometric ratio from the equation:

moles of B = moles of A × ratio B/A

Worked example

25.0 cm^3 of a solution of sulfuric acid of concentration 0.0565 mol dm^{-3} was neutralised by 23.4 cm^3 of a solution of sodium hydroxide. Calculate the concentration of the sodium hydroxide solution.

Equation: $2NaOH + H_2SO_4 \rightarrow Na_2SO_4 + 2H_2O$

Step 1: amount of H_2SO_4 = 0.0565 × 25.0/1000 = 0.0014125 mol

Step 2: amount of NaOH = 0.0014125 × 2/1 = 0.002825 mol

Step 3: concentration of NaOH = 0.002825 mol/0.0234 dm^3 = 0.121 mol dm^{-3}

In step 2, the stoichiometric ratio is 2/1 as there are 2NaOH to $1H_2SO_4$ in the equation.

Checklist

Before attempting the questions on this topic, check that you know:

- ☐ the physical properties of the elements
- ☐ the reactions of the elements with oxygen, chlorine and water
- ☐ the reactions of their oxides with water and acid
- ☐ the reactions of their hydroxides with acid
- ☐ the trend in ionisation energies within a group
- ☐ the trend in solubilities of group 2 sulfates and hydroxides
- ☐ the reason for the trend in thermal stability of group 1 and group 2 nitrates and carbonates
- ☐ the flame colours of their compounds
- ☐ how to do calculations based on titration data

Testing your knowledge and understanding

For the first set of questions, cover the margin, write your answers, then check to see if you are correct.

★ Which group 2 metal has the highest melting temperature?

★ Which is the most soluble group 2 hydroxide?

★ Barium compounds are normally poisonous, but barium sulfate is given to people in order to outline their gut in radiography (X-ray imaging). Why is barium sulfate not poisonous?

★ Name the products of heating:
a lithium nitrate **b** sodium nitrate

Answers

Beryllium

Barium hydroxide

It is extremely insoluble.

a Lithium oxide, nitrogen dioxide and oxygen
b Sodium nitrite and oxygen

The **answers** to the **numbered questions** are on page 133

1 You are given unlabelled solid samples of lithium chloride, potassium chloride and barium chloride. Describe the tests that you would do to find out which was which.

2 Explain why sodium compounds give a yellow colour to a flame.

3 Explain why the first ionisation energy of magnesium is greater than the first ionisation energy of calcium.

4 Write balanced equations for the reactions of:
a calcium oxide with water
b magnesium hydroxide with dilute hydrochloric acid

5 Write balanced equations for the reactions of
a calcium with oxygen
b calcium with water
c magnesium with chlorine
d magnesium with steam

6 Explain why the addition of dilute sodium hydroxide to a solution of magnesium chloride produces a white precipitate, but no precipitate forms when dilute sodium hydroxide is added to a solution of barium chloride.

7 State which group 2 element forms the least thermally stable carbonate. Explain why this is the case.

8 Write balanced equations for the thermal decomposition of the following compounds. If there is no reaction at temperatures that can be achieved in a laboratory, give this as your answer.

a lithium nitrate, sodium nitrate and magnesium nitrate

b sodium carbonate, magnesium carbonate and barium carbonate

9 1.33 g of sodium carbonate, Na_2CO_3, was weighed out and dissolved in water. The solution was made up to 250 cm^3. A burette was filled with this solution. 25.0 cm^3 of a solution of hydrochloric acid was pipetted into a conical flask and titrated with the sodium carbonate solution. The mean titre was 23.75 cm^3.

$$Na_2CO_3 + 2HCl \longrightarrow 2NaCl + H_2O + CO_2$$

Calculate:

a the amount (in moles) of sodium carbonate weighed out

b the concentration of the sodium carbonate solution

c the amount (in moles) of sodium carbonate in the titre

d the amount (in moles) of hydrochloric acid in 25.0 cm^3

e the concentration of the hydrochloric acid solution

10 2.41 g of a solid acid of formula H_2X was weighed out, dissolved and made up to 250 cm^3 with distilled water. 25.0 cm^3 samples were titrated against 0.200 mol dm^{-3} sodium hydroxide solution. The mean titre was 26.80 cm^3.

$$H_2X + 2NaOH \longrightarrow Na_2X + 2H_2O$$

Calculate:

a the amount (in moles) of sodium hydroxide in the mean titre

b the amount (in moles) of acid H_2X in 25.0 cm^3

c the amount (in moles) of acid in 250 cm^3

d the molar mass of the acid H_2X

Topic **2.4**

The periodic table: group 7 (chlorine to iodine)

Introduction

★ Halogens are oxidising agents. Their strength decreases down the group:
  - Chlorine is the strongest oxidising agent and iodine is the weakest.
  - Chloride ions are the weakest reducing agents and are, therefore, the hardest to oxidise; iodide ions are the strongest reducing agents and are, therefore, the easiest to oxidise.

★ They form halide ions (such as Cl^-) and oxoanions (such as ClO_3^-).

★ Solutions of the hydrogen halides are strong acids.

Things to learn

Physical properties of the elements

	Physical state	Colour of solution in water	Colour of solution in hexane
Chlorine	Green gas	Pale green	Pale green
Bromine	Red-brown liquid	Red-brown	Red-brown
Iodine	Dark grey shiny solid	Brown	Violet

Reaction with metals

Halogens react with metals to form halides, for example:

$2Fe + 3Cl_2 \longrightarrow 2FeCl_3$

$Fe + I_2 \longrightarrow FeI_2$

As chlorine is the stronger oxidising agent, iron(III) chloride is formed; iodine forms iron(II) iodide.

Reaction with non-metals

Halogens react with non metals to form covalent halides, for example:

$2P + 5Cl_2 \longrightarrow 2PCl_5$

Reactions of halides with aqueous silver ions

This is used as a test for ionic halides.

In solution with aqueous silver ions:

- ★ a chloride gives a white precipitate of AgCl that dissolves in dilute ammonia
- ★ a bromide gives a cream coloured precipitate that is insoluble in dilute ammonia but dissolves in concentrated ammonia.
- ★ an iodide gives a yellow precipitate that is insoluble in both dilute and concentrated ammonia

Things to understand

Redox reactions

- ★ Chlorine is a powerful oxidising agent and is reduced to the -1 state. The half-equation is:

  $Cl_2 + 2e^- \longrightarrow 2Cl^-$

  Other halogens react similarly but they are weaker oxidising agents in the order $Cl_2 > Br_2 > I_2$.

- ★ Iodides are strong reducing agents:

  $2I^- \longrightarrow I_2 + 2e^-$

Reaction of halides with concentrated sulfuric acid

The gaseous hydrogen halide is first produced and this may be oxidised by the concentrated sulfuric acid.

- ★ Chlorides produce steamy acid fumes of HCl:

  $KCl + H_2SO_4 \longrightarrow HCl + KHSO_4$

- ★ Bromides produce steamy acid fumes of HBr with some red-brown bromine:

  $KBr + H_2SO_4 \longrightarrow HBr + KHSO_4$

  $2KBr + 3H_2SO_4 \longrightarrow 2KHSO_4 + Br_2 + SO_2 + 2H_2O$

- ★ Iodides give clouds of violet iodine vapour.

HBr is just powerful enough as a reducing agent to reduce some of the concentrated sulfuric acid and itself be oxidised to bromine. The HI initially produced is a very powerful reducing agent and is oxidised to iodine.

Reaction of halides with other halogens

- ★ Chlorine displaces bromine and iodine from bromides and iodides:

  $Cl_2 + 2KBr \longrightarrow Br_2 + 2KCl$

  $Cl_2 + 2KI \longrightarrow I_2 + 2KCl$

- ★ Bromine displaces iodine from iodides:

  $Br_2 + 2KI \longrightarrow I_2 + 2KBr$

Reaction of halogens and halides with iron ions

- ★ Chlorine and bromine oxidise aqueous iron(II) ions:

  $Cl_2 + 2Fe^{2+} \longrightarrow 2Cl^- + 2Fe^{3+}$

  $Br_2 + 2Fe^{2+} \longrightarrow 2Br^- + 2Fe^{3+}$

- ★ Iodide ions will reduce iron(III) ions

  $2I^- + 2Fe^{3+} \longrightarrow I_2 + 2Fe^{2+}$

These reactions show that the strength as oxidising agents decreases in the order $Cl_2 > Br_2 > I_2$.

Disproportionation

This occurs when an element in a single species is simultaneously oxidised and reduced. It follows that there must be at least two atoms of the element, with the same oxidation number, on the left-hand side of the equation and that the element must be able to exist in at least three different oxidation states.

★ Chlorine disproportionates in cold alkali to the –1 and +1 states:

$$\underset{0}{Cl_2} + 2OH^- \longrightarrow \underset{-1}{Cl^-} + \underset{+1}{OCl^-} + H_2O$$

★ In hot alkali, it disproportionates into the –1 and +5 states:

$$3\underset{0}{Cl_2} + 6OH^- \longrightarrow 5\underset{-1}{Cl^-} + \underset{+5}{ClO_3^-} + 3H_2O$$

Hydrogen halides as acids

★ All hydrogen halides are strong acids (except for hydrogen fluoride, HF, which is a weak acid) that dissolve with water to form H^+ ions:

$$HCl(g) + aq \longrightarrow H^+(aq) + Cl^-(aq)$$

★ All react with ammonia to form a solid ammonium salt as white fumes:

$$HBr(g) + NH_3(g) \longrightarrow NH_4Br(s)$$

Iodine titrations

The concentration of an oxidising agent can be measured by reacting it with excess acidified potassium iodide and then titrating the liberated iodine against sodium thiosulfate, $Na_2S_2O_3$. The thiosulfate is added from the burette until the iodine colour fades to a pale straw colour. Starch is then added and the titration continued until the dark blue starch–iodine complex is decolorised.

$$2Na_2S_2O_3 + I_2 \longrightarrow Na_2S_4O_6 + 2NaI$$

1 mol of iodine reacts with 2 mol thiosulfate.

Worked example

Excess potassium iodide was added to 25.0 cm^3 portions of a solution containing Fe^{3+} ions. The liberated iodine was titrated against 0.123 mol dm^{-3} sodium thiosulfate solution and the mean titre was 24.40 cm^3. Calculate the concentration of iron(III) ions in the solution.

$$2Fe^{3+} + 2I^- \longrightarrow I_2 + 2Fe^{2+}$$

$$2Na_2S_2O_3 + I_2 \longrightarrow Na_2S_4O_6 + 2NaI$$

Answer

amount (moles) of thiosulfate in titre = 0.123 × 0.02440 = 0.00300 mol

amount (moles) of iodine reacted = 0.5 × 0.00300 = 0.00150 mol

amount (moles) of Fe^{3+} ions in 25.0 cm^3 = 2 × 0.00150 = 0.00300 mol

concentration of Fe^{3+} ions = moles/volume in dm^3 = 0.00300/0.0250

= 0.120 mol dm^{-3}

Checklist

Before attempting questions on this topic, check that you can recall

- ☐ the physical properties of the elements
- ☐ the reactions of halogens with metals and non-metals

- ☐ the tests for the halides
- ☐ the reactions of concentrated sulfuric acid with halides
- ☐ the relative strengths of the halogens as oxidising agents
- ☐ examples of −1, 0, +1 and +5 oxidation states of chlorine and the disproportionation of chlorine
- ☐ iodine titration calculations

Testing your knowledge and understanding

For the first set of questions, cover the margin, write your answers, then check to see if you are correct.

★ Name the halogen that is a liquid at room temperature.

★ What is the product of the reaction between

a zinc and bromine?

b iron and iodine?

★ Which of Cl_2, Br_2 and I_2 is the strongest oxidising agent?

★ Which of HCl, HBr and HI is the strongest reducing agent?

★ What is the colour of the precipitate obtained when aqueous bromide and silver ions are mixed?

★ What would you see when aqueous chlorine is added to a solution of potassium bromide?

Answers

Bromine

a $ZnBr_2$

b FeI_2

Chlorine

Hydrogen iodide

Cream

A red-brown colour

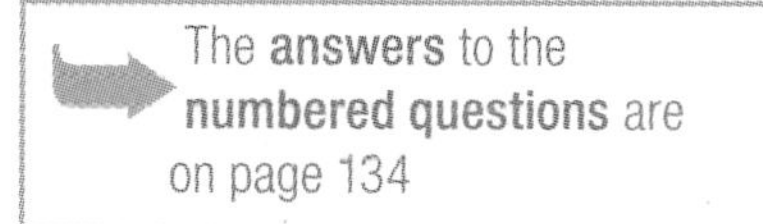
The **answers** to the **numbered questions** are on page 134

1 It is thought that a sample of solid sodium carbonate has been contaminated by sodium chloride. How would you test for the presence of chloride ions in this sample?

2 Explain why, when concentrated sulfuric acid is added to solid sodium chloride, hydrogen chloride gas is produced and chlorine is not, whereas when concentrated sulfuric acid is added to solid sodium iodide the products are iodine and just a trace of hydrogen iodide.

3 Define disproportionation and give an example of a disproportionation reaction of either chlorine or a chlorine compound.

4 The purity of a sample of potassium iodate(V), KIO_3, was found by first dissolving 1.00 g of impure potassium iodate(V) in water and making the solution up to 250 cm^3 with distilled water. 25.0 cm^3 portions were taken and excess potassium iodide and dilute sulfuric acid were added. The liberated iodine was titrated with 0.105 mol dm^{-3} sodium thiosulfate solution. The mean titre was 22.75 cm^3.

$$IO_3^- + 5I^- + 6H^+ \rightarrow 3I_2 + 3H_2O$$
$$I_2 + 2S_2O_3^{2-} \rightarrow 2I^- + S_4O_6^{2-}$$

Calculate:

a the amount (in moles) of thiosulfate ions in the mean titre

b the amount (in moles) of liberated iodine

c the amount (in moles) of iodate(V) ions in 25.0 cm^3

d the mass of pure potassium iodate(V) in 250 cm^3 of solution

e the percentage purity of the sample of potassium iodate(V)

Topic **2.5**

Kinetics

Introduction

The rate of a reaction is determined by:

★ the frequency at which the molecules collide

★ the fraction of the colliding molecules that possess enough kinetic energy 'to overcome the activation energy barrier'

★ the orientation of the molecules on collision

Things to learn

★ **Activation energy** is the minimum energy that the reactant molecules must have when they collide in order for them to form product molecules.

★ Factors that control the rate of a reaction are:
- the concentration of a reactant in a solution
- the pressure, if the reactants are gases
- the temperature
- the presence of a catalyst
- the surface area of any solid reactants
- light, for photochemical reactions

Things to understand

Collision theory

★ The effect of an increase in concentration of a solution, or an increase in pressure of a gas, is to increase the frequency of collision of the molecules and hence increase the rate of reaction.

★ The effect of heating a gas or a solution is to make the molecules or ions move faster and, therefore, to have a greater average kinetic energy. This increases the fraction of colliding molecules with a combined energy equal to or greater than the activation energy, which results in a greater proportion of successful collisions.

This can be shown by the Maxwell–Boltzmann distribution (see Figure 2.4) of molecular energies at two temperatures T_1 and T_2 with $T_2 > T_1$.

An increase in temperature also causes an increase in the rate of collision. However, this is not significant in comparison with the increase in the fraction of molecules with energy equal to or greater than the activation energy.

Note that the peak has moved to the right for T_2 and that the areas under the curves are the same.

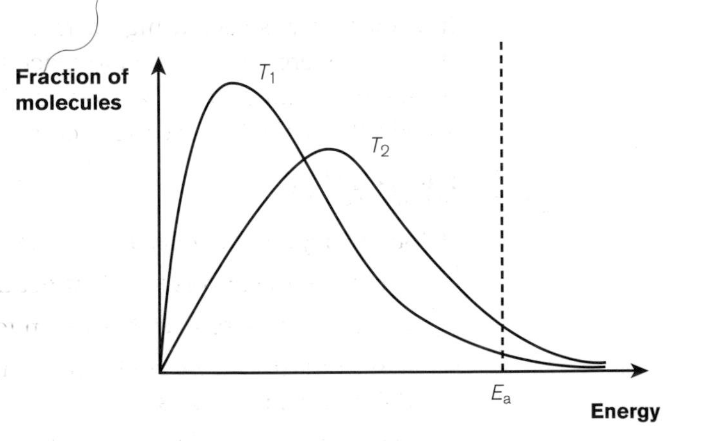

Figure 2.4 Maxwell–Boltzmann distribution

★ A **catalyst** works by providing an alternative route with a lower activation energy. Thus, at a given temperature, a greater proportion of the colliding molecules will possess the lower activation energy of the catalysed route and so the reaction will be faster. This is shown in Figure 2.5.

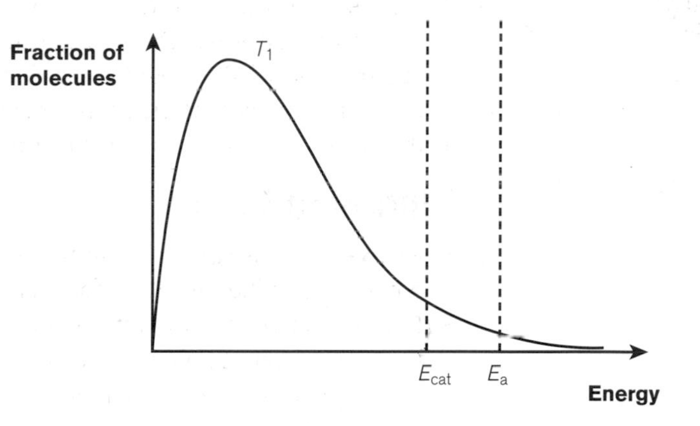

Figure 2.5 Effect of a catalyst

The energy-profile diagrams for an uncatalysed and a catalysed reaction are shown in Figure 2.6.

Do not say that a catalyst lowers the activation energy. Note that the overall ΔH values are the same for both paths.

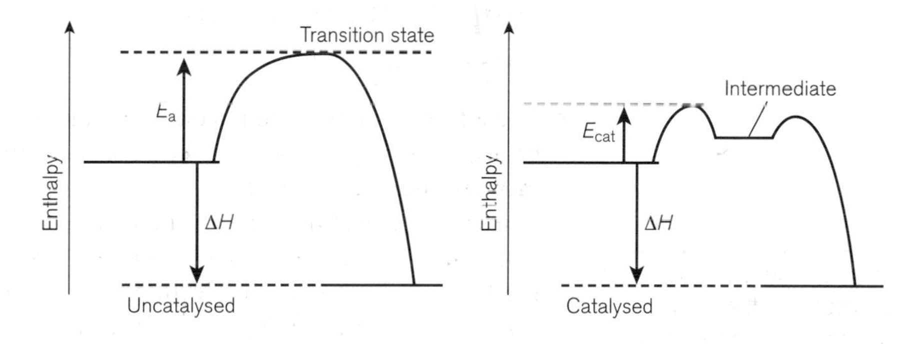

Figure 2.6 Energy-profile diagrams for an uncatalysed and a catalysed reaction

Kinetic stability

If a reaction has such a high activation energy that almost no molecules possess sufficient energy on collision to react, the system is said to be kinetically stable. An example is a mixture of petrol and air, which is thermodynamically unstable but kinetically stable. No reaction occurs unless the mixture is ignited.

Checklist

Before attempting the questions on this topic, check that you:

- ☐ can recall the factors which affect the rate of reaction
- ☐ can explain these rate changes in terms of collision theory
- ☐ can draw the Maxwell–Boltzmann distribution of molecular energies at two different temperatures
- ☐ can use this to explain the effect of both a change in temperature and the addition of a catalyst
- ☐ understand the concept of kinetic stability

Testing your knowledge and understanding

For the following questions, cover the margin, write your answers, then check to see if you are correct.

★ For a reaction involving gases, state three factors that control the rate of the reaction.

★ For a reaction carried out in solution, state three factors that affect the rate of reaction.

★ For the reaction of a solid with a gas or with a solution, state one other factor that affects the rate of reaction.

Answers
Pressure, temperature, and catalyst
Concentration, temperature and presence of a catalyst
The surface area of the solid

The **answers** to the **numbered questions** are on page 134

1 Define:
  a activation energy b catalyst

2 Explain, in terms of collision theory, why the three factors that you have chosen above control the rate of a reaction in the gaseous phase.

3 a Draw the Maxwell–Boltzmann distribution of energy curve for a gas
  (i) at room temperature (label this T_1) and
  (ii) at 50°C (label this T_2)
  c Mark in a typical value for the activation energy of a reaction that proceeds steadily at room temperature

4 Explain, in terms of activation energy, why animal products such as meat and milk stay fresher when refrigerated.

5 Draw energy profile diagrams of:
  a an exothermic reaction occurring in a single step
  b the same reaction in the presence of a suitable catalyst
  c a reaction in which the reactants in an endothermic reaction are kinetically stable
  d a reaction in which the reactants in an exothermic reaction are kinetically unstable

Topic **2.6**

Chemical equilibria

Introduction

Many reactions do not go to completion because they are reversible. As the rate of a reaction is dependent on the concentration of the reactants, the reaction will proceed up to the point at which the rate of the forward reaction equals the rate of the reverse reaction, and so there is no further change in concentrations. The system is then said to be at equilibrium.

Things to learn and understand

Dynamic equilibrium

★ At equilibrium, the rate of the forward reaction equals the rate of the reverse reaction.

★ Both products and reactants are being formed and consumed continuously but their concentrations do not change.

★ This can be demonstrated by the use of isotopes. For example, for the reaction:

$$CH_3COOH(l) + C_2H_5OH(l) \rightleftharpoons CH_3COOC_2H_5(l) + H_2O(l)$$

the four substances are mixed in their equilibrium concentrations; the water is labelled with the isotope ^{18}O. Using a mass spectrometer, it is found that the ^{18}O isotope eventually appears in both the ethanoic acid and the ethanol — the concentrations of the four substances do not change.

Effect of changes in conditions on position of equilibrium

Le Chatelier's principle states that when one of the factors governing the position of equilibrium is changed, the position alters in such a way as to restore the original conditions.

★ Le Chatelier's principle may help you to *predict* the direction of the change in equilibrium position, but it does not *explain* it.

★ **Temperature:** an *increase* in temperature moves the position of equilibrium in the *endo*thermic direction; a *decrease* in temperature moves the equilibrium position in the *exo*thermic direction.

$$N_2(g) + 3H_2(g) \rightleftharpoons 2NH_3(g) \qquad \Delta H = -92.4 \text{ kJ mol}^{-1}$$

As this reaction is exothermic left to right, an increase in temperature will result in a reduction in the amount of ammonia, thus lowering the yield.

Do not say 'the side with a smaller volume' because a gas always fills its container.

★ **Pressure:** this applies only to reactions involving gases. An increase in pressure drives the equilibrium to the side with fewer gas molecules. Thus, for the reaction above, an increase in pressure will result in more ammonia in the equilibrium mixture, i.e. an increased yield. This is because there are only two gas molecules on the right-hand side of the equation and four on the left.

★ **Concentration:** this applies mainly to equilibrium reactions in solution. If a substance is physically or chemically removed from an equilibrium system, the equilibrium shifts to produce more of that substance. For example, for the reaction:

$$2CrO_4^{2-}(aq) + 2H^+(aq) \rightleftharpoons Cr_2O_7^{2-}(aq) + H_2O(l)$$

addition of alkali will remove the H^+ ions, so the equilibrium will move to the left.

★ **Catalyst:** this has *no* effect on the position of equilibrium. What it does is to *increase the rate at which equilibrium is reached*. Thus, a catalyst allows a reaction to be carried out at a reasonable rate at a lower temperature.

Methane hydrate is found deep underwater in cold seas. When brought to the surface, the pressure decreases and the temperature increases.

Worked example

Methane hydrate, $[CH_4(H_2O)_6]$, decomposes endothermically into methane and water:

$$[CH_4(H_2O)_6](s) \rightleftharpoons CH_4(g) + 6H_2O(l)$$

What is the effect on solid methane hydrate of:

a reducing the pressure

b increasing the temperature

Answers

a As there is 1 gas molecule on the right-hand side and none on the left, reduction in pressure will drive the equilibrium to the right.

b The reaction is endothermic left to right, so an increase in temperature will cause the position of equilibrium to move to the right.

Experimental observations

Effect of adding acid to aqueous chromate ions, $CrO_4^{2-}(aq)$

* The equilibrium is:

$$2CrO_4^{2-}(aq) + H_2O(l) \rightleftharpoons Cr_2O_7^{2-}(aq) + 2OH^-(aq)$$

yellow — orange

* The yellow solution turns orange as acid is added. This is because the addition of acid removes OH^- ions and so drives the equilibrium to the right.

Effect of compressing nitrogen dioxide, NO_2

* The equilibrium is:

$$2NO_2(g) \rightleftharpoons N_2O_4(g)$$

brown — colourless

If the equilibrium did *not* shift, the colour would darken because the NO_2 would become more concentrated.

* The brown colour fades as the pressure is increased. This is because the increase in pressure drives the equilibrium to the side with fewer gas molecules, which is the right-hand side in this example.

Checklist

Before attempting the questions on this topic, check that you:

- [ ] understand that equilibria are dynamic
- [ ] can deduce the effect of changes in temperature, pressure and concentration on the position of equilibrium
- [ ] can explain observed changes in terms of the shift in the position of equilibrium

Testing your knowledge and understanding

For the following questions, cover the margin, write your answers, then check to see if you are correct.

* Consider the endothermic reaction:

$$CH_4(g) + H_2O(g) \rightleftharpoons CO(g) + 3H_2(g)$$

Give the effect (if any) of:

a an increase in temperature on
 (i) the rate of the reaction
 (ii) the position of equilibrium

b an increase in pressure on the position of equilibrium

c the addition of more catalyst on the position of equilibrium

Answers

a (i) it increases
(ii) it moves to the right
b it moves to the left
c none

The **answers** to the **numbered questions** are on page 135

1 Consider the reversible reaction at equilibrium:

$$A(g) \rightleftharpoons B(g)$$

Which statement(s) is/are true about this system?

a There is no further change in the concentrations of A or B.
b No reactions occur, now that the system has reached equilibrium.
c The rate of formation of B is equal to the rate of formation of A.

2 Consider the equilibrium reaction:

$$PCl_5(g) \rightleftharpoons Cl_2(g) + PCl_3(g) \qquad \Delta H = +93\,kJ\,mol^{-1}$$

State and explain the effect on the position of equilibrium of:

a decreasing the temperature

b halving the volume of the container
c adding a catalyst

3 Ethanoic acid and ethanol react reversibly:
$CH_3COOH + C_2H_5OH \rightleftharpoons CH_3COOC_2H_5 + H_2O \qquad \Delta H = 0\,kJ\,mol^{-1}$
a Explain the effect of adding an alkali
b What will happen to the position of equilibrium if the temperature is increased from 25°C to 35°C?

4 White insoluble lead(II) chloride reacts reversibly with aqueous chloride ions to form a colourless solution:
$PbCl_2(s) + 2Cl^-(aq) \rightleftharpoons PbCl_4^{2-}(aq)$
State and explain what you would see when concentrated hydrochloric acid is added to the equilibrium mixture.

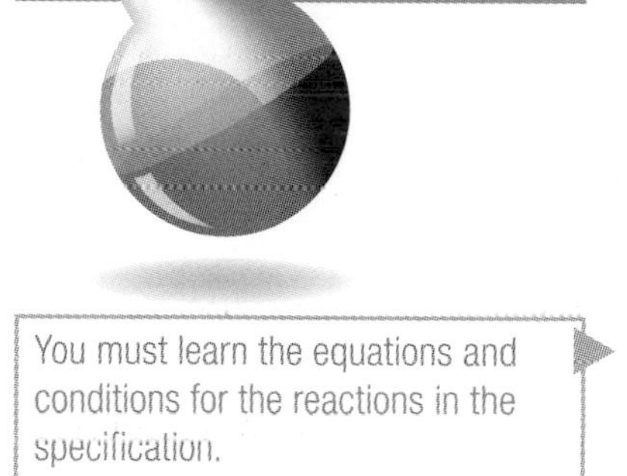

Topic 2.7

Organic chemistry

Introduction

You must learn the equations and conditions for the reactions in the specification.

★ An organic compound consists of a chain of one or more carbon atoms and may also contain one or more functional groups. A functional group gives a compound certain chemical properties. For instance, the $-CH_2OH$ group reacts in a similar way in all compounds. Thus, knowledge of the chemistry of ethanol, CH_3CH_2OH, enables you to predict the reactions of all compounds containing the $-CH_2OH$ group.

Substance	Alkene	Alcohol	Aldehyde	Ketone	Acid	Ester	Nitrile	Amine
Group	>C=C<	–C–OH	H(R)C=O	R(R')C=O	–C(=O)OH	R–C(=O)O–R'	–C≡N	–C–NH$_2$

Things to learn

When you write structural formulae, check that each carbon atom has four bonds, each oxygen has two bonds and each hydrogen and each halogen has only one bond.

★ **Structural formula:** an unambiguous formula, showing each group separately with any double bonds clearly shown, e.g. propane can be written as $CH_3CH_2CH_3$ and propan-2-ol as $CH_3CH(OH)CH_3$.
★ **Skeletal formula:** lines represent C–C and double bonds; no hydrogen atoms are shown on the chain:

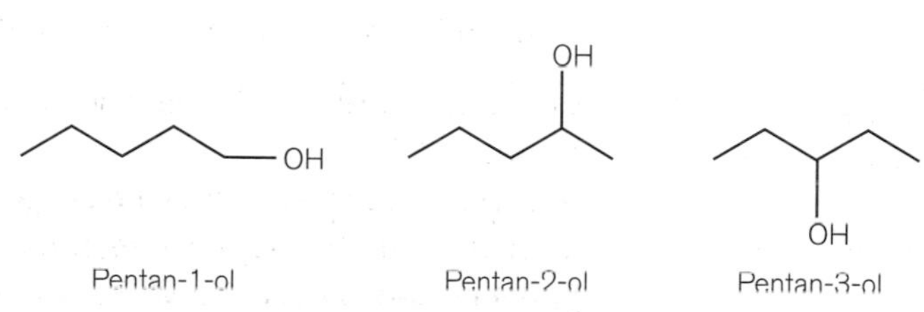

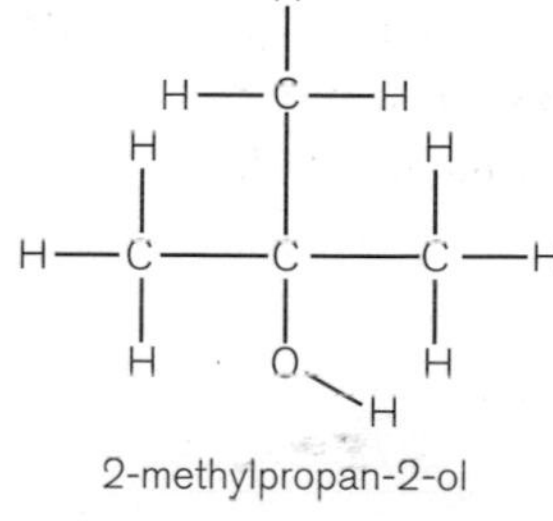
2-methylpropan-2-ol

★ **Displayed formula:** shows all the atoms and all the bonds separately. For example, the displayed formula of 2-methylpropan-2-ol is shown opposite:
★ **Isomers:** two or more compounds with the same molecular formula.
★ **Hydrolysis:** a reaction (often catalysed by aqueous acid or aqueous alkali) in which water splits an organic molecule into two other molecules.

★ **Nucleophile:** a species that is attracted to a positive centre and has a lone pair of electrons, which it donates to form a new covalent bond.

Things to understand

Alcohols

★ Primary (1°) alcohols contain a $-CH_2OH$ group.
★ Secondary (2°) alcohols have two carbon atoms attached to a >CHOH group
★ Tertiary (3°) alcohols have three carbon atoms attached to a COH group.

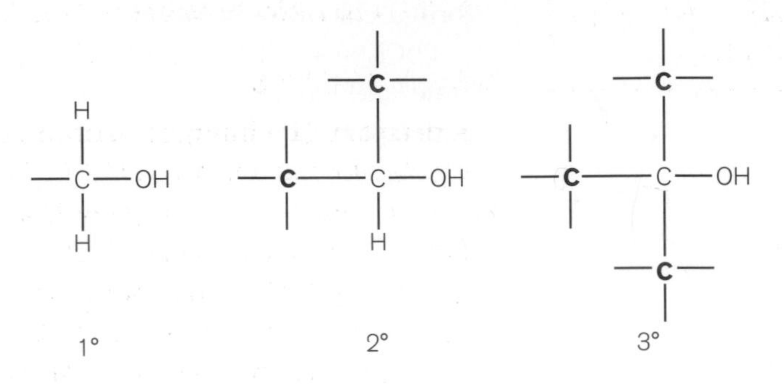

Figure 2.7 Primary, secondary and tertiary alcohols

Reactions of alcohols

★ **Combustion:** all alcohols burn with a clean flame in excess air to produce carbon dioxide and water, for example with ethanol:

$C_2H_5OH + 3O_2 \rightarrow 2CO_2 + 3H_2O$

> Bioethanol is a renewable fuel that can be made from sugar or grain.

★ **Reaction with sodium:** all alcohols react with sodium to form hydrogen gas and a white solid. With ethanol the solid formed is sodium ethoxide:

$C_2H_5OH + Na \rightarrow C_2H_5O^-Na^+ + \frac{1}{2}H_2$

> This is a test for an –OH group in alcohols *and* acids. Steamy acidic fumes are given off when phosphorus pentachloride is added to the dry organic substance.

★ **Halogenation:**

a All alcohols react with phosphorus pentachloride, PCl_5, to form **chloroalkanes**, for example:
Equation: $C_2H_5OH + PCl_5 \rightarrow C_2H_5Cl + HCl + POCl_3$
Conditions: room temperature; the alcohol must be dry

b Alcohols react with solid sodium bromide and 50% sulfuric acid to give **bromoalkanes**, for example:
Equations: $NaBr + H_2SO_4 \rightarrow HBr + NaHSO_4$
then $HBr + C_2H_5OH \rightarrow C_2H_5Br + H_2O$
Conditions: add sulfuric acid mixed with the alcohol to solid sodium bromide at room temperature

> Concentrated sulfuric acid must not be used when trying to make a bromoalkane, because it will oxidise the HBr to bromine. Even 50% sulfuric will oxidise HI.

c Alcohols react with phosphorus and iodine to give **iodoalkanes**, for example:
Equations: $2P + 3I_2 \rightarrow 2PI_3$
then $3C_2H_5OH + PI_3 \rightarrow 3C_2H_5I + H_3PO_3$
Conditions: add the alcohol to a mixture of moist red phosphorus and iodine at room temperature

★ **Oxidation:** with aqueous orange potassium dichromate(VI), $K_2Cr_2O_7$, acidified with dilute sulfuric acid

a A **primary alcohol** is oxidised via an aldehyde to a carboxylic acid. The solution turns green as Cr^{3+} ions are formed:
Equations: $CH_3CH_2OH + [O] \rightarrow CH_3CHO + H_2O$
then $CH_3CHO + [O] \rightarrow CH_3COOH$

An aldehyde is not hydrogen bonded and so has a lower boiling point than the hydrogen-bonded alcohol. This means that it boils off from the hot solution, leaving the alcohol behind.

Hint

In equations in organic chemistry, an oxidising agent can be written as [O] and a reducing agent as [H], but the equation must still balance.

Conditions: to stop at the aldehyde — add potassium dichromate(VI) in dilute sulfuric acid to the hot alcohol and distil off the aldehyde. The temperature must be between the boiling points of the alcohol and the aldehyde.

To prepare the acid — boil the alcohol under reflux with excess acidified oxidising agent.

b A **secondary alcohol** is oxidised to a ketone. The solution turns green.
Equation: $CH_3CH(OH)CH_3 + [O] \rightarrow CH_3COCH_3 + H_2O$
Conditions: boil the alcohol and the acidified oxidising agent under reflux.

c **Tertiary alcohols** are *not* oxidised and so do *not* turn the solution green.

Halogenoalkanes

- A **primary (1°) halogenoalkane** contains the CH_2X group where X is a halogen, e.g. 1-chlorobutane, $CH_3CH_2CH_2CH_2Cl$.
- A **secondary (2°) halogenoalkane** has two carbon atoms attached to the CHX group, e.g. 2-chlorobutane, $CH_3CH_2CHClCH_3$.
- A **tertiary (3°) halogenoalkane** has three carbons attached to the CX group, e.g. 2-chloro-2-methylpropane, $(CH_3)_3CCl$.
- A halogenoalkane contains a δ+ carbon atom and so is attacked by nucleophiles. They do not hydrogen bond and so have lower boiling points than similar alcohols. They cannot form hydrogen bonds with water and are, therefore, insoluble.

Preparation

They can be prepared from alkenes (see page 25) or alcohols (see above).

Reactions

Hint

The mechanism is described on page 58.

2-iodopropane, CH_3CHICH_3, is used as the example.

- With *aqueous* potassium (or sodium) hydroxide: undergoes a nucleophilic substitution reaction forming an alcohol
  Equation: $CH_3CHICH_3 + KOH \rightarrow CH_3CH(OH)CH_3 + KI$
  Conditions: Boil under reflux with aqueous potassium (or sodium) hydroxide
- With *ethanolic* potassium hydroxide: undergoes an elimination reaction forming an alkene
  Equation: $CH_3CHICH_3 + KOH \rightarrow CH_3CH{=}CH_2 + KI + H_2O$
  Conditions: Boil under reflux with a concentrated solution of potassium hydroxide in ethanol
- With aqueous silver nitrate: water hydrolyses the C–X bond and X^- ions are formed. These react with Ag^+ ions to form a precipitate of AgX.
  Equation: $CH_3CHICH_3(l) + H_2O(l) \rightarrow CH_3CH(OH)CH_3(aq) + H^+(aq) + I^-(aq)$
  then $I^-(aq) + Ag^+(aq) \rightarrow AgI(s)$
  The rate of reaction is C–Cl < C–Br < C–I and 1° < 2° < 3°.
- With ammonia: an amine is formed
  Equation: $CH_3CHICH_3 + 2NH_3 \rightarrow CH_3CH(NH_2)CH_3 + NH_4I$
  Conditions: heat excess concentrated ammonia in ethanol with the halogenoalkane in a sealed tube

The rates of all these reactions increase from C–Cl to C–Br to C–I, because the bond enthalpy and hence the bond strength decreases from C–Cl to C–Br to C–I. The weaker the bond the lower is the activation energy and hence the faster the reaction.

Test for a halogenoalkane

1. Warm the substance with aqueous sodium hydroxide and a few drops of ethanol (to increase solubility)
2. Add excess dilute nitric acid
3. Add silver nitrate solution

Result:

- ★ White precipitate, soluble in dilute ammonia indicates a chloro-compound
- ★ Cream precipitate, soluble in concentrated ammonia indicates a bromo-compound
- ★ Yellow precipitate, insoluble in concentrated ammonia indicates an iodo-compound.

Uses of halogenoalkanes

CFCs makes excellent refrigerants. However, they are so stable to air and water that they diffuse into the stratosphere where they are broken down by light energy to form chlorine radicals (see page 24).

Summary of reactions

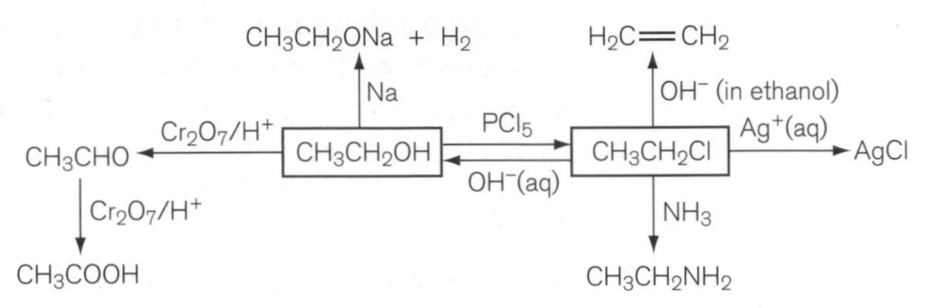

Checklist

Before attempting the questions on this topic, check that you:

- ☐ can name simple alcohols and halogenoalkanes
- ☐ can draw structural formulae of structural and geometric isomers
- ☐ know the reaction of alcohols with sodium, and their oxidation and halogenation reactions
- ☐ know the reactions of halogenoalkanes with potassium hydroxide (both aqueous and ethanolic), aqueous silver nitrate and ammonia

Testing your knowledge and understanding

For the first set of questions, cover the margin, write your answers, then check to see if you are correct.

- ★ Name $CH_3CH_2CH(CH_3)CHClCH_2OH$
- ★ Name the organic product of reacting propan-2-ol with:
  - **a** acidified potassium dichromate
  - **b** KBr and 50% sulfuric acid
- ★ What would you observe when a piece of sodium is added to a little ethanol?
- ★ Name the organic product of reacting 2-bromo-2-methylpropane with:
  - **a** aqueous potassium hydroxide
  - **b** potassium hydroxide dissolved in ethanol
  - **c** ammonia dissolved in ethanol
- ★ A compound X contained 54.5% carbon, 36.4% oxygen and 9.1% hydrogen by mass. Calculate its empirical formula.

Answers

2-chloro-3-methylpentan-1-ol

a propanone
b 2-bromopropane

Bubbles of gas and a white solid

a 2-methylpropan-2-ol
b 2-methylpropene
c 2-amino-2-methylpropane

C_2H_4O

The **answers** to the **numbered questions** are on page 135

1 Write out the structural formulae of the structural isomers of
 a C_3H_8O (alcohols only)
 b C_4H_9Cl

2 Write equations and name the products obtained when 2-bromopropane is:
 a warmed with aqueous dilute sodium hydroxide
 b heated under reflux with a concentrated solution of potassium hydroxide in ethanol
 c heated in a sealed tube with a concentrated solution of ammonia in ethanol

3 Write the structural formula of the organic product formed, if any, give its name and say what you would see when:
 a propan-1-ol is heated under reflux with dilute sulfuric acid and excess aqueous potassium dichromate(VI)
 b 2-methylpropan-2-ol is heated under reflux with dilute sulfuric acid and aqueous potassium dichromate(VI)
 c phosphorus pentachloride is added to propan-2-ol

4 Why does 1-bromopropane react more slowly with water than 1-iodopropane does?

5 When 5.67 g of cyclohexene, C_6H_{10}, reacted with excess bromine, 15.4 g of $C_6H_{10}Br_2$ was obtained. Calculate the theoretical yield of $C_6H_{10}Br_2$ and hence the percentage yield of the reaction.

Topic 2.8

Mechanisms

Introduction

★ A mechanism shows the movement of electrons during the various steps of a reaction and also the intermediate radicals or ions in that reaction.
★ Reactants are classified as either electrophiles, nucleophiles or radicals because of the way they behave in a mechanism.

Things to learn

★ **Homolytic fission**: a bond breaks and one electron goes to each atom, forming radicals
★ **Heterolytic fission**: a bond breaks and the two electrons go to one atom, forming charged species that can then be attacked by electrophiles or nucleophiles
★ **Addition**: a reaction in which two molecules react to form a single molecule
★ **Substitution**: a reaction in which an atom or group of atoms in one molecule is replaced by another atom or group of atoms
★ **Elimination:** a reaction in which the elements of a simple molecule, for example HBr or H_2O, are removed from an organic molecule and are not replaced by any other atom or group of atoms
★ **Nucleophile**: a species that seeks out positive centres; it has a lone pair of electrons which it donates to form a new covalent bond
★ **Electrophile**: a species that seeks out negative centres and accepts a lone pair of electrons to form a new covalent bond
★ A **curly arrow** represents the movement of a *pair* of electrons.
★ A **curly half-arrow** represents the movement of a *single* electron.

Things to understand

Bonding and reactivity

★ A C–I bond is weaker than a C–Cl bond. Therefore, iodoalkanes are hydrolysed by aqueous sodium hydroxide at a faster rate than chloroalkanes

★ δ+ sites, such as the δ+ carbon atom in a halogenoalkane, are attacked by nucleophiles. Electron-rich areas, as in the π-bond in alkenes, are attacked by electrophiles.

Mechanism of nucleophilic substitution

The mechanism is called S_N2 because the reaction is a substitution (S), the attacking reagent is a nucleophile (N) and there are two species involved in the first step.

★ The carbon atom in the C–X group is δ+ and is attacked by the lone pair of electrons on a nucleophile

★ In the reaction between chloroethane and aqueous hydroxide ions, a new σ-bond forms between the negative oxygen and the δ+ carbon and at the same time the σ-bond between the carbon and the chlorine breaks:

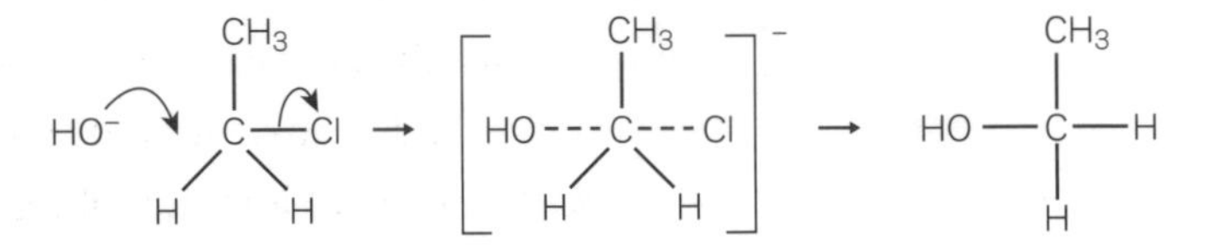

A carbocation is an ion with a positive charge on a carbon atom.

★ Some secondary and all tertiary halogenoalkanes react by a S_N1 mechanism. The first step is the ionisation of the C–X bond to form X^- and a carbocation. This is then attacked by the nucleophile:

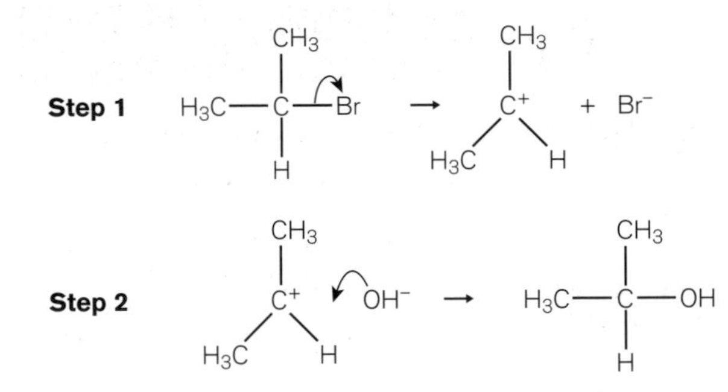

Mechanism of electrophilic addition

See page 26.

This is a chain reaction as, in each propagation step, one radical gives rise to another radical.

Mechanism of free radical substitution

See page 24–25.

Ozone depletion

Oxygen and ozone (O_3) are in equilibrium in the stratosphere. Ozone is formed by some O_2 molecules absorbing UV radiation at a wavelength of 200 nm.

$$O_2 \xrightarrow{UV} 2O\bullet$$

then $O\bullet + O_2 \rightarrow O_3$

This ozone is decomposed by UV radiation but at a longer wavelength

$$O_3 \xrightarrow{UV} O_2 + O\bullet$$

then $O\bullet + O\bullet \rightarrow O_2$

Certain man-made substances, such as CFCs from aerosols and refrigerants, and oxides of nitrogen from aeroplane engines, upset this equilibrium.

The nitric oxide, NO, reacts with ozone:

$NO + O_3 \rightarrow NO_2 + O_2$

The nitrogen(IV) oxide produced then reacts with oxygen radicals to form nitrogen(II) oxide, which can then destroy more ozone:

$NO_2 + O\bullet \rightarrow NO + O_2$

Checklist

Before attempting the questions on this topic, check that you can:

- [ ] differentiate between homolytic and heterolytic fission
- [ ] define a nucleophile and an electrophile
- [ ] recognise δ+ and electron-rich sites in organic molecules
- [ ] write the mechanism for the:
  - nucleophilic substitution of halogenoalkanes
  - electrophilic addition to alkenes
  - free radical substitution of alkanes
  - depletion of the ozone layer by nitrogen oxides

Testing your knowledge and understanding

For the first set of questions, cover the margin, write your answers, then check to see if you are correct.

★ Identify which of the following is a free radical, a nucleophile and a carbocation:
NH_3, CH_3^+ and $CH_3\bullet$

★ Which bond is weaker, C–Cl or C–I?

★ Which bond is more polar, C–Cl or C–I?

★ Which reacts faster with nucleophiles, C_2H_5Cl or C_2H_5I?

★ Which reacts faster with nucleophiles, $CH_3CH_2CH_2Cl$ or $CH_3CHClCH_3$?

1 Give the S_N1 mechanism for the reaction between 2-iodopropane and aqueous hydroxide ions.

2 Give the S_N2 mechanism for the reaction between iodomethane and aqueous hydroxide ions.

3 Give the mechanism for the free radical reaction between ethane and bromine in the presence of UV light. Your answer should show the initiation step together with appropriate curly arrows, two propagation steps and one termination step.

4 Give the mechanism for the electrophilic addition reaction between propene and hydrogen bromide. Why is the product you have shown more likely to be formed than its isomer?

Answers

$CH_3\bullet$ is a free radical, NH_3 is a nucleophile and CH_3^+ is a carbocation.

C–I

C–Cl

C_2H_5I

$CH_3CHClCH_3$

The **answers** to the **numbered questions** are on page 135

Topic **2.9**

Mass spectra and IR

Introduction

The key requirements of this topic are to be able to:

- ⋆ interpret mass spectra, including molecular ion and fragment peaks
- ⋆ use infrared spectra to identify functional groups in organic compounds
- ⋆ explain why some molecules do not absorb in the infrared region

Things to learn

- ⋆ In **mass spectra** the peaks are caused by positively charged ions.
- ⋆ The peak with the highest *m/e* ratio is caused by the molecular ion.
- ⋆ In **infrared spectra** the peaks are caused by the bending or stretching of covalent bonds.

Things to understand

Mass spectra

- ⋆ You must first identify the molecular ion peak.
- ⋆ The peaks due to fragments are at particular values lower than that of the molecular ion peak.

Common fragments are shown in the table below.

Fragment lost	*m/e* of peak due to fragment ion	*m/e* of peak due to ion from remainder of the molecule
CH_3	15	(M – 15)
C_2H_5 or CHO	29	(M – 29)
CH_2OH	31	(M – 31)
COOH	45	(M – 45)
Cl	35 and 37	(M – 35) and (M – 37)

Worked example

A compound was thought to be either propanal, CH_3CH_2CHO, or propanone, CH_3COCH_3. It had peaks in its mass spectrum at 58, 43, 29 and 15. Identify the unknown and state the identities of the species causing these lines.

Answer

The line at 29 is caused by either a CHO^+ ion or a $C_2H_5^+$ ion. Only propanal has these groups, so the unknown is propanal. The peaks are due to:

at 58, the molecular ion, $(CH_3CH_2CHO)^+$
at 43, the $(CH_2CHO)^+$ ion (the molecule having lost CH_3)
at 29, the fragments $(CHO)^+$ and $(C_2H_5)^+$
at 15, the fragment $(CH_3)^+$

Infrared spectra

- ⋆ You are not expected to know the wavenumbers due to absorption by different functional groups, but you must be able to use data, such as those in the table below, to assign functional groups to peaks.

Wavenumber/cm^{-1}	Bond	Functional group
3000–2850	C–H	Alkanes
3100–3000	C–H	Alkenes
3600–3200	O–H	Alcohols (hydrogen bonded)
3700–3600	O–H	Alcohols (not hydrogen bonded)
3300–2500	O–H	Carboxylic acids
3500–3300	N–H	Amines
1740–1720	C=O	Aldehydes
1730–1700	C=O	Ketones
1725–1700	C=O	Carboxylic acids
800– 600	C–Cl	Chloroalkanes
600– 500	C–Br	Bromoalkanes
About 500	C–I	Iodoalkanes

The peaks due to hydrogen-bonded O–H and N–H are very broad.

- **Following a reaction** — the progress of an organic reaction can be followed by observing one peak decreasing as another increases. The oxidation of propan-2-ol to propanone can be followed in this way. The broad peak at around 3300 cm^{-1} decreases as the sharp peak at around 1710 cm^{-1} increases.
- **Conditions for IR absorption** — there must be a change of dipole moment for the bending or stretching to be infrared active. Thus, only molecules with polar bonds can absorb IR radiation.
  - Non-polar molecules with polar bonds can absorb if the bending or stretching causes a change in dipole moment. Carbon dioxide is a linear molecule, so the symmetrical stretching of the two C=O bonds does not cause a change in dipole moment and it is, therefore, IR inactive. However, the asymmetrical stretch and the bending vibrations result in a change in dipole moment and so are IR active.
  - The same applies to methane, which has polar C–H bonds but is a symmetrical molecule and so is not polar. Some of its vibrations cause the dipole to change from zero and so those vibrations are IR active.
  - Water has polar bonds and the bond angle is 104.5°; it is bent molecule so it is polar. Its vibrations are IR active.

Greenhouse gases

Some or all of the vibrations of CO_2, CH_4, H_2O and ozone, O_3, are IR active, so these gases in the air will absorb infrared radiation. This means that they absorb some of the IR radiation which is being radiated from the Earth. Thus they are greenhouse gases.

Oxygen and nitrogen do not have any polar bonds and so cannot absorb IR radiation.

Checklist

Before attempting the questions on this topic, check that you can:

- ☐ identify the ions responsible for the peaks in a mass spectrum
- ☐ assign peaks in an IR spectrum to different functional groups
- ☐ indicate how IR spectroscopy can be used to follow a reaction

Testing your knowledge and understanding

Answers
$(CH_2COOH)^+$ and $(COOH)^+]$
A C=C group
Because it does not contain any polar bonds
Because it contains polar bonds

The **answers** to the **numbered questions** are on page 136

For the first set of questions, cover the margin, write your answers, then check to see if you are correct.

★ What species are responsible for the peaks at *m/e* of 59 and at *m/e* of 45 in the mass spectrum of propanoic acid, CH_3CH_2COOH?

★ A neutral organic compound absorbs infrared radiation at 3050 cm^{-1}. What functional group does it contain?

★ Why is chlorine, Cl_2, not a greenhouse gas?

★ Why is sulfur hexafluoride, SF_6, a greenhouse gas?

1 Bromine has a relative atomic mass of 79.9 and consists of two isotopes ^{79}Br and ^{81}Br. A compound X of molecular formula C_3H_7Br (molar mass 123 g mol^{-1}) had peaks in its mass spectrum at *m/e* values of 124, 122 and 29, but none at a *m/e* value of 123.

- **a** What species are responsible for the peaks at *m/e* values of 124, 122 and 29?
- **b** Explain why there is no peak at *m/e* = 123.
- **c** Write the structural formula of the organic compound X.
- **d** What will be the relative heights of the peaks at *m/e* 122 and 124?

2 Explain which of the following will absorb infrared radiation and which will not:

a NO **b** Cl_2 **c** CCl_4

3 Ethanoic acid and propan-1-ol both have a molar mass of 60.0 g mol^{-1}. Examine the IR spectrum below.

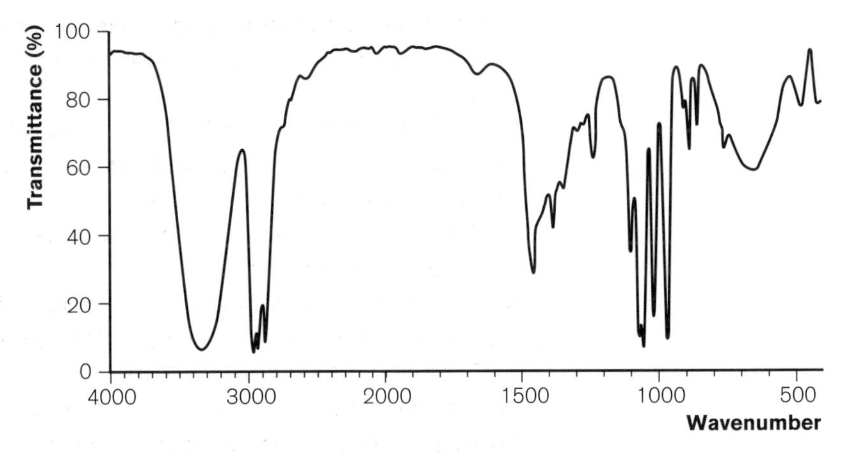

- **a** State whether this is the spectrum of ethanoic acid or propan-1-ol. Justify your answer.
- **b** What bonds caused the peaks at 3333 cm^{-1} and 2963 cm^{-1}? (You may use the table of absorption frequencies on page 61.)

4 The rate of the reaction:

$$C_2H_5Cl + 2NH_3 \rightarrow C_2H_5NH_2 + NH_4Cl$$

was studied using IR spectroscopy. Give the frequency of the peak in the IR spectrum of the reaction mixture that would decrease and that of the peak which would increase as the reaction proceeded. Justify your answer.

5 What is a greenhouse gas?

Topic 2.10

Green chemistry

Introduction

The key to this topic is to understand how industry is changing to lessen its impact on the environment.

Things to learn

- ★ The **carbon footprint** of a fuel is the total mass of carbon dioxide released into the environment by the production *and* combustion of that fuel.
- ★ **Carbon neutrality** is when the carbon footprint equals the amount of carbon dioxide absorbed by photosynthesis.
- ★ Atom economy can be worked out according to the following formulae:

$$\textbf{atom economy} = \frac{\textbf{mass of atoms in the product}}{\textbf{mass of the atoms in all the reactants}} \times \textbf{100}$$

or

$$\frac{\textbf{molar mass of product} \times \textbf{number of formula units of it in the equation}}{\textbf{sum of molar masses of all the reactants}} \times \textbf{100}$$

- ★ **Anthropological climate change** is the effect on the climate caused by the activities of mankind, such as burning fossil fuels.

Things to understand

- ★ Use of **renewable resources** — for example recycling aluminium cans, rather than extracting aluminium from bauxite ore.
- ★ Use of **less hazardous materials** — dangerous pesticides such as DDT and organophosphates are being replaced by alternatives that are similar to natural plant pesticides.
- ★ Use of **reactions with higher atom economies** — ethanoic acid can be manufactured by the fermentation of glucose to ethanol, followed by bacterial oxidation. The overall reaction is:

  $C_6H_{12}O_6 + 2O_2 \rightarrow 2CH_3COOH + 2CO_2 + 2H_2O$

  The atom economy is 49%.

  A new method of producing ethanoic acid is the reaction between carbon monoxide and methanol using an iridium complex ion as a catalyst:

  $CO + CH_3OH \rightarrow CH_3COOH$

  The atom economy in this process is 100%.
- ★ **More efficient use of energy** — the pharmaceutical industry uses microwaves to heat up aqueous reaction mixtures. The heat losses by this method are considerably less than those using electric heaters, Bunsen burners or other flame heaters.

The most abundant greenhouse gas is water vapour, which contributes about 95% of the total greenhouse effect.

- ★ **Effects of different greenhouse gases** — the main anthropogenic greenhouse gases are carbon dioxide, methane, oxides of nitrogen and CFCs. The latter two are more powerful absorbers of infrared radiation per mole; carbon dioxide is the least powerful. However, there is nearly ten times as much carbon dioxide as methane and only small amounts of the other gases, so carbon dioxide has most effect on global warming.
- ★ **Carbon footprint of fuels** — hydrogen and biofuels do not have a zero carbon footprint. Hydrogen has to be manufactured, either using electricity or by the

reaction between methane and steam. Both processes are responsible for large quantities of carbon dioxide. Bioethanol has about the same carbon footprint as petrol when the footprint of growing the crop, harvesting and purification are taken into account. Biodiesel also has a significant carbon footprint caused by the fuels used in growing and transesterification.

- **CFCs and the ozone layer** — CFCs, such as CF_2Cl_2, destroy the ozone layer in a chain reaction, thus a few CFC molecules destroy a large number of ozone molecules. The CFCs are stable to normal atmospheric reactions, but are decomposed at high altitudes to give chlorine radicals.

$$CF_2Cl_2 \xrightarrow{UV} CF_2Cl\bullet + Cl\bullet$$

The chlorine radicals then deplete the ozone layer:

$$Cl\bullet + O_3 \rightarrow O_2 + ClO\bullet$$

The oxygen radicals are produced by the decomposition of ozone by UV light: $O_3 + UV \rightarrow O\bullet + O_2$

The chlorine radicals are reformed by the reaction between ClO• radicals and O• radicals:

$$ClO\bullet + O\bullet \rightarrow Cl\bullet + O_2$$

Checklist

Before attempting the questions on this topic, check that you can:

- ☐ define atom economy
- ☐ understand the concept of carbon footprint
- ☐ discuss the carbon footprint of different fuels
- ☐ discuss the relative effect of different greenhouse gases
- ☐ explain how CFCs damage the ozone layer

Testing your knowledge and understanding

For the first set of questions, cover the margin, write your answers, then check to see if you are correct.

- What is the atom economy for the production of ethyl ethanoate from ethanol and ethanoic acid?

$$C_2H_5OH + CH_3COOH \rightarrow CH_3COOC_2H_5 + H_2O$$

- Carbon dioxide, methane, nitrogen oxide and CFCs are all greenhouse gases.
  - a Which has the least impact in terms of heat absorbed per mole?
  - b Which has the greatest impact on global warming?

Answers

83%

a Carbon dioxide

b Carbon dioxide

1 Chloroethane can be produced by reacting ethanol with phosphorus pentachloride:

$$C_2H_5OH + PCl_5 \rightarrow C_2H_5Cl + POCl_3 + HCl$$

- a Calculate the atom economy of this reaction.
- b Chloroethane can also be made by the catalytic dehydration of ethanol followed by the addition of hydrogen chloride. Comment about the relative merits, in terms of atom economy, of these two methods.

The **answers** to the **numbered questions** are on page 136

2 Study the table below and explain which of the gases would be a greenhouse gas if released into the atmosphere and which would not.

	Difference in electronegativity	Shape	Dipole moment/D
CH_4	0.4	Tetrahedral	0
CI_4	0	Tetrahedral	0
H_2	0	Linear	0
CO	1.0	Linear	0.10

3 Explain why bioethanol used as a fuel is not carbon neutral.

4 Explain how the damage to the ozone layer by CFCs, such as $CF_2ClCFCl_2$, is due to a chain reaction.

Practice Unit Test 2

Section A

1 The oxidation number of oxygen in KO_2 is

A −2 B −1 C −½ D +2

2 The oxidation number of oxygen in the O_2^{2-} ion is

A −2 B −1 C −½ D +2

3 The oxidation number of oxygen in OF_2 is

A −2 B −1 C −½ D +2

4 Which of the following is *not* a disproportionation reaction?

A $3NaOCl \rightarrow 2NaCl + NaClO_3$
B $2CuCl \rightarrow Cu + CuCl_2$
C $Cl^- + OCl^- + H_2O \rightarrow Cl_2 + 2OH^-$
D $4ClO_3^- \rightarrow 3ClO_4^- + Cl^-$

5 The shape of an ammonia molecule is

A tetrahedral B triangular pyramidal
C square planar D V-shaped

6 Which of the following substances *cannot* form intermolecular hydrogen bonds?

A fluoromethane, CH_3F
B hydrogen fluoride, HF
C ethanol, CH_3CH_2OH
D ethanoic acid, CH_3COOH

7 The strongest intermolecular forces between hydrogen chloride molecules are

A covalent bonds
B hydrogen bonds
C permanent dipole forces
D induced dipole (dispersion) forces

8 The F–S–F bond angle in SF_6 is

A 60° B 90° C 109.5° D 120°

9 The F–S–F bond angle in SF_2 is

A 105° B 109.5° C 120° D 180°

10 The group 1 carbonate that is most easily decomposed on heating is

A Li_2CO_3 B Na_2CO_3 C K_2CO_3 D Rb_2CO_3

11 Which of the following ions has the smallest radius?

A F^- B Cl^- C Na^+ D K^+

12 An organic compound of molecular formula $C_4H_{10}O$ gave steamy fumes when reacted with phosphorus pentachloride. Its mass spectrum had a peak at *m/e* = 74 due to the molecular ion and at *m/e* values of 15 and 31 due to fragment ions, but no peak at *m/e* = 29. It is

A butan-1-ol, $CH_3CH_2CH_2CH_2OH$
B butan-2-ol, $CH_3CH(OH)CH_2CH_3$
C 2-methylpropan-2-ol, $(CH_3)_3COH$
D 2-methylpropan-1-ol, $(CH_3)_2CHCH_2OH$

13 Which of the following is *not* a nucleophile?

A H_2O B OH^- C NH_3 D NH_4^+

14 When hydrogen bromide adds to propene, the major product is 2-bromopropane because

A a secondary halogenoalkane is more stable than a primary halogenoalkane
B the secondary intermediate is more stable than the primary intermediate
C bromine is more electronegative than hydrogen
D hydrogen is more electronegative than bromine

15 Methane reacts with excess chlorine in the presence of UV light by a free radical mechanism to form a mixture of mono and dichloromethane. Which of the following is *not* present in the chain reaction

A $CH_3\bullet$ **B** $H\bullet$ **C** $Cl\bullet$ **D** $CH_2Cl\bullet$

16 Aqueous hydrochloric acid reacts with magnesium in an exothermic reaction. Which set of conditions will give the fastest reaction with identical 2 cm strips of magnesium?

A 0.2 mol dm^{-3} acid and a temperature of 25°C
B 0.1 mol dm^{-3} acid and a temperature of 75°C
C 0.1 mol dm^{-3} acid and a temperature of 25°C
D 0.2 mol dm^{-3} acid and a temperature of 20°C

17 A white solid gave rise to a lilac colour in a flame test and its solution gave a cream precipitate with aqueous silver nitrate. The precipitate dissolved in concentrated ammonia. The solid is

A KBr **B** KI **C** LiCl **D** LiBr

18 Chloroethane can be prepared by various reactions. Which of the following has the highest atom economy?

A $C_2H_5OH + PCl_5 \rightarrow C_2H_5Cl + POCl_3 + HCl$
B $3C_2H_5OH + PCl_3 \rightarrow 3C_2H_5Cl + H_3PO_3$
C $C_2H_4 + HCl \rightarrow C_2H_5Cl$
D $C_2H_5Br + HCl \rightarrow C_2H_5Cl + HBr$

19 Nitrogen tetroxide is in equilibrium with nitrogen dioxide:

$N_2O_4(g) \rightleftharpoons 2NO_2(g)$ $\Delta H = +58\ kJ\ mol^{-1}$
colourless brown

Which of the following statements is *not* true?

A When heated the equilibrium mixture becomes paler
B When heated the equilibrium mixture becomes darker
C When the pressure is increased the equilibrium mixture gets paler
D When the pressure is increased the equilibrium mixture gets warmer

20 Lithium nitrate decomposes on heating. Which of the equations below is correct for this reaction?

A $LiNO_3 \rightarrow LiNO_2 + ½O_2$
B $2LiNO_3 \rightarrow Li_2O + 2NO_2 + ½O_2$
C $Li(NO_3)_2 \rightarrow Li(NO_2)_2 + O_2$
D $Li(NO_3)_2 \rightarrow LiO + 2NO_2 + ½O_2$

Total: 20 marks

Section B

21 Explain the following statements:

a Ammonia, NH_3, has a higher boiling point than phosphine, PH_3. **(3 marks)**
b Hydrogen chloride, HCl, has a lower boiling point than hydrogen bromide, HBr. **(2 marks)**
c Chloromethane, CH_3Cl, has a higher boiling point than propane, C_3H_8. **(2 marks)**
d It is easier to decompose magnesium carbonate than calcium carbonate. **(2 marks)**
e Iodoethane, C_2H_5I, reacts faster than chloroethane, C_2H_5Cl, with aqueous sodium hydroxide. **(2 marks)**

Total: 11 marks

22 This question is about sulfuric acid.

a The critical reaction in the manufacture of sulfuric acid is:

$2SO_2(g) + O_2(g) \rightleftharpoons 2SO_3(g)$ $\Delta H = -196\ kJ\ mol^{-1}$

Explain, in terms of equilibrium and kinetics, why a temperature of 425°C is used rather than a higher or lower temperature. **(3 marks)**

b 1.56 g of concentrated sulfuric acid was carefully dissolved in water and made up to 250 cm^3 with distilled water. 25.0 cm^3 samples were titrated against 0.111 mol dm^{-3} sodium hydroxide solution. The mean titre was 27.45 cm^3.

$2NaOH + H_2SO_4 \rightarrow Na_2SO_4 + 2H_2O$

Calculate:

(i) the amount (in moles) of sodium hydroxide in the titre **(1 mark)**

(ii) the amount (in moles) of H_2SO_4 in 25.0 cm^3 of solution and hence in the original sample. **(2 marks)**
(iii) the mass of H_2SO_4 in the sample. **(1 mark)**
(iv) the percentage composition of H_2SO_4 in the concentrated sulfuric acid. **(1 mark)**

c What would be seen, apart from effervescence, when concentrated sulfuric acid is added to:
(i) solid potassium chloride? **(1 mark)**
(ii) solid potassium bromide? **(2 marks)**

d Write equations for the reactions in **c (ii)** **(2 marks)**

e Use your answers to **c** to explain why a mixture of concentrated sulfuric acid and potassium bromide is not used in the preparation of bromoalkanes from alcohols. **(1 mark)**

Total: 14 marks

23 a (i) What would you see when sodium is added to a small amount of ethanol? **(2 marks)**
(ii) Write the equation for this reaction **(1 mark)**

b What would you see if the following were heated under reflux with acidified potassium dichromate(VI)?
(i) butan-2-ol **(1 mark)**
(ii) 2-methyl-propan-2-ol **(1 mark)**

c Name the organic product when propan-1-ol is heated under reflux with acidified potassium dichromate(VI) **(1 mark)**

d Draw the apparatus that you would use in **c** **(3 marks)**

Total: 9 marks

24 a Write equations for the reaction of bromoethane with the following:
(i) aqueous potassium hydroxide **(1 mark)**
(ii) potassium hydroxide dissolved in ethanol **(1 mark)**

b Cyanide ions, CN^-, are nucleophiles and react with halogenoalkanes in the same way as do hydroxide ions. Suggest a mechanism for the reaction of cyanide ions and bromoethane. **(3 marks)**

c Describe the tests and observations that would show that an organic liquid is a bromoalkane. **(5 marks)**

Total: 10 marks

Section C

25 a Draw dot-and-cross diagrams for:
(i) water **(1 mark)**
(ii) carbon dioxide **(1 mark)**

b (i) Explain why water molecules are V-shaped and carbon dioxide molecules are linear. **(3 marks)**
(ii) Explain why water is polar and carbon dioxide is non-polar. **(2 marks)**

c Explain how carbon dioxide can absorb infrared radiation and hence be a greenhouse gas. **(2 marks)**

d Discuss the relative effects of carbon dioxide and methane on global warming. **(2 marks)**

e Explain, in terms of the effect on the environment, why the scientific community recommended that the manufacture of CFCs should be banned. You must illustrate your answer with equations. **(5 marks)**

Total: 16 marks

Paper Total: 80 marks

UNIT 4 Rates, equilibria and further organic chemistry

Topic 4.1 Rates: how fast?

Introduction

You must revise Unit 2 Kinetics (pages 48–50), particularly how the changes in the Maxwell–Boltzmann distribution of energies with an increase in temperature affect the rate of reaction.

Things to learn

- ★ **The rate constant, *k***, is the constant of proportionality in a rate equation. Its value depends on the activation energy of the reaction and the temperature. A reaction with a high activation energy has a low value of *k* and a slow rate.

Order of reaction	Units of *k*
Zero order	s^{-1}
First order	$mol^{-1}\,dm^3\,s^{-1}$
Second order	$mol^{-2}\,dm^6\,s^{-1}$

- ★ The **order with respect to one substance** (the partial order) is the power to which the *concentration* of that substance is raised in the rate equation.
- ★ The **order of reaction** is the sum of the partial orders.
- ★ The **activation energy, E_a**, is the total kinetic energy that the molecules must have on collision in order for them to be able to react.
- ★ **Half-life, $t_{1/2}$**, is the time taken for the concentration to fall from any selected value to half that value.
- ★ A **heterogeneous catalyst** is in a different phase from the reactants; a **homogeneous catalyst** is in the same phase.
- ★ The **rate-determining step** is the slowest step in a mechanism.

The half-life of a first-order reaction is constant.

A solid catalyst is in a different phase from gaseous or aqueous reactants.

Things to understand

Measurement of rate

- ★ The concentration of a reactant or product is measured after a period of time. The average rate = change in concentration/time elapsed.
- ★ If a gas is produced, the average rate is proportional to volume of gas/time elapsed (or mass lost/time elapsed).
- ★ If the time is measured either for all of one reactant (e.g. a strip of magnesium) to react completely or for enough product to be formed to hide a mark on the apparatus (e.g. enough sulfur to hide a cross), the rate is proportional to 1/time elapsed.

Worked example

In a reaction between 0.100 mol dm^{-3} sodium hydroxide and excess 2-bromopropane, the time for the solution to become neutral was 35 s. Calculate the rate of reaction.

Answer

concentration of sodium hydroxide changes from 0.100 mol dm^{-3} to zero in 35 s

so, rate = (0.100 – 0)/35 = 0.0029 mol dm^{-3} s^{-1}

Determination of rate equation and rate constant from rates

Consider the reaction:

A + B ⟶ products

The rates of reactions with different initial concentrations of A and B are found and by comparing pairs of experiments the order with respect to each substance can be found.

Look for two experiments where the concentration of only *one* substance varies. If doubling that concentration causes the rate to double, the order with respect to that substance is 1.

Experiment	[A]/mol dm^{-3}	[B]/mol dm^{-3}	Rate/mol dm^{-3} s^{-1}
1	0.1	0.1	0.001
2	0.2	0.1	0.002
3	0.1	0.2	0.004

- From experiments 1 and 2: [A] doubles, [B] is constant and the rate doubles. Therefore, the order with respect to A is 1.
- From experiments 1 and 3: [B] doubles, [A] is constant and the rate increases four-fold. Therefore, the order with respect to B is 2.

rate of reaction = $k[A]^1[B]^2$, and the reaction is third order

rate constant, $k = \frac{\text{rate}}{[A][B]^2} = \frac{0.001}{0.1 \times 0.1^2} = 1\ \text{mol}^{-1}\ \text{dm}^3\ \text{s}^{-1}$

Experimental techniques for following a reaction

Titration method

Mix the chemicals and start the clock. At intervals withdraw a sample and add it to iced water to slow the reaction Then titrate one of the substances in the reaction. This method can be used if an acid, an alkali or iodine is a reactant or a product.

Colorimetric method

This can be used when either a reactant or a product is coloured (iodine or potassium manganate(VII) are examples). The colorimeter must first be calibrated using solutions of the coloured substance of known concentrations. Then the reactants are mixed and the clock started. The intensity of the colour is measured as a function of time. The concentration of the coloured substance is proportional to the amount of light absorbed.

Iodine clock reaction

The oxidation reaction of iodide ions by an oxidising agent such as hydrogen peroxide can be followed by a 'clock' method:

- ★ Known volumes of hydrogen peroxide solution, dilute sulfuric acid and sodium thiosulfate solution are mixed with a few drops of starch. A known volume of potassium iodide is added, the solution stirred and a clock started.
- ★ As iodine is produced by the oxidation of iodide ions, it reacts with the thiosulfate ions. When all the thiosulfate has reacted, the iodine then produced turns the starch blue.
- ★ At this point the clock is stopped.
- ★ The rate of reaction is proportional to the volume of thiosulfate/time.

Graphs of results

If the half-lives are constant the reaction is first order. If they increase, the reaction is second order.

If the concentration of a reactant is plotted against time, three shapes of graph are likely depending on the order of the reaction. If the graph is a straight line, the rate of reaction is constant and so the reaction is first order. If the graph is curved, measure two consecutive half-lives.

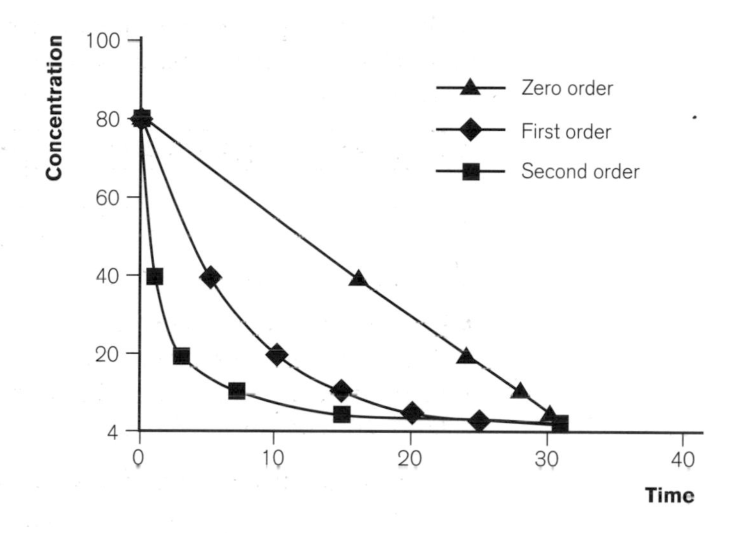

Activation energy

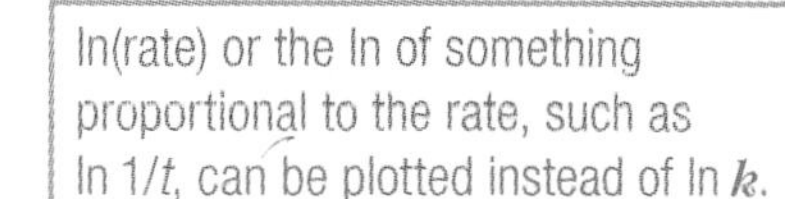

The activation energy can be found by plotting a graph of ln k against $1/T$, where T is the temperature in kelvin. The gradient of this line is $-E_a/R$, where $R = 8.31\ \mathrm{J\ mol^{-1}\ K^{-1}}$.

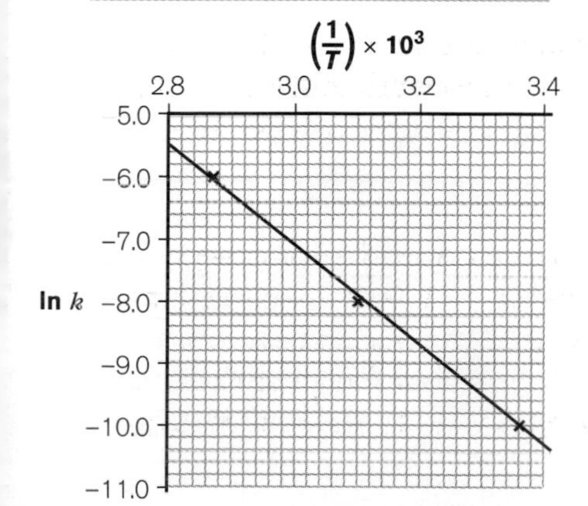

k	ln k	Temperature/°C	Temperature/K	1/temperature/K^{-1}
0.0000454	-10.0	25	298	0.00336
0.000335	-8.0	50	323	0.00310
0.00248	-6.0	75	348	0.00287

$$\text{gradient} = \frac{-10 - (-6.00)}{(0.00336 - 0.00287)}$$

$$= \frac{-4}{0.00049} = -8163$$

$$E_a = -\text{gradient} \times R = -(-8163 \times 8.31) = +66400\ \mathrm{J\ mol^{-1}} = +66.4\ \mathrm{kJ\ mol^{-1}}$$

Mechanisms

A suggested mechanism must be consistent with the orders of reaction. The partial order of any species that occurs in the mechanism *after* the rate-determining step will be zero. The rate-determining step is the slowest step.

The hydrolysis of halogenoalkanes can take place either by a S_N1 or a S_N2 mechanism (see page 55).

- rate = *k*[halogenoalkane]
  The mechanism is S_N1 and the nucleophile reacts in the second step, which is faster than the rare-determining step.
- rate = *k*[halogenoalkane][OH^-]
  The mechanism is S_N2 and both the halogenoalkane and the nucleophile are involved in the first step, which is the rate-determining step.

Checklist

Before attempting the questions on this topic, check that you:

- [ ] can define rate constant, order of reaction and half-life
- [ ] can deduce rate equations from rate of reaction data
- [ ] can suggest suitable methods for following reactions
- [ ] can deduce the order of a reaction from concentration–time graphs
- [ ] calculate the activation energy from rate constant data
- [ ] understand that information about mechanisms can be deduced from the partial orders of the reactants

Testing your knowledge and understanding

For the following set of questions, cover the margin, write your answers, then check to see if you are correct.

- The following results were obtained from a study of the reaction:
  $NO_2(g) + CO(g) \rightarrow NO(g) + CO_2(g)$

Experiment	[NO_2]/mol dm^{-3}	[CO]/mol dm^{-3}	Relative rate
1	0.02	0.02	1
2	0.04	0.02	4
3	0.02	0.04	1

  a What is the order with respect to NO_2 and with respect to CO?
  b Determine the order of reaction and write down the rate equation.

Answers

a From experiments 1 and 2, order with respect to NO_2 is 2.
From experiments 1 and 3, order with respect to CO is 0.

b Order of reaction is 2 + 0 = 2; rate = $k[NO_2]^2$

- The decomposition of N_2O_5 is first order. At 200°C, the reaction has a half-life of 25 minutes. Calculate how long it will take for the concentration of N_2O_5 to fall to 6.25% of its original value.

100% to 6.25% takes four half-lives. Therefore, time taken = 4 × 25 = 100 minutes

The **answers** to the **numbered questions** are on page 139

1 The rate of the second order reaction:
$2HI(g) \rightarrow H_2(g) + I_2(g)$
is 2.0×10^{-4} mol dm^{-3} s^{-1} when [HI] = 0.050 mol dm^{-3} at 785 K. Calculate the value of the rate constant and give its units.

2 Describe how you would follow the rate of the following reaction at 60°C:
$CH_3COOH(aq) + CH_3OH(aq) \rightarrow CH_3COOCH_3(l) + H_2O(l)$.

3 The decomposition of 3-oxobutanoic acid, CH_3COCH_2COOH, was studied.

$$CH_3COCH_2COOH \rightarrow CH_3COCH_3 + CO_2$$

The results, at 40°C, are shown in the table below.

Time/min	0	26	52	78
[3-oxobutanoic acid]/mol dm^{-3}	1.6	0.80	0.40	0.20

Deduce the order of the reaction.

4 The kinetics of a reaction were studied at different temperatures. Use the graph to calculate the gradient and hence the activation energy.

(R = 8.3 J K^{-1} mol^{-1}; $\ln k = \ln A - E_a/RT$)

$\left(\frac{1}{T}\right)$ /K^{-1}

0.0028 0.0030 0.0032 0.0034

ln k

−5.0 −6.0 −7.0 −8.0 −9.0 −10.0 −11.0

Topic 4.2 Entropy: how far?

Introduction

★ This section indicates whether or not a reaction is likely to take place.

Things to learn

★ **Entropy,** symbol S, is a measure of disorder. Its value depends on molecular complexity and the physical state of a substance. Its units are J K^{-1} mol^{-1}.

★ $S_{solid} < S_{liquid} < S_{gas}$

★ Three important relationships are:

$$\Delta S_{system} \text{ for a chemical change} = \Sigma S_{products} - \Sigma S_{reactants}$$

$$\Delta S_{surroundings} = -\Delta H/T$$

$$\Delta S_{total} = \Delta S_{system} + \Delta S_{surroundings} = \Delta S_{system} - \Delta H/T$$

★ **Lattice energy** is the energy change when 1 mol of an ionic solid is formed from its gaseous ions.

★ **Enthalpy of hydration** is the enthalpy change when 1 mol of a gaseous ion is dissolved in excess water.

Hint

Beware: entropy values are in joules and enthalpy values in kilojoules, so ΔH must be multiplied by 1000 before being divided by the temperature, which must be in kelvin.

Things to understand

Feasibility of reaction

A reaction that is feasible may not occur if the activation energy is too high. If so, the reaction is said to be kinetically inert.

- ★ A reaction is feasible (thermodynamically favourable) if the value of ΔS_{total} is positive.
- ★ Endothermic reactions: ΔH is positive and hence $\Delta S_{surroundings}$ is negative. ΔS_{system} must be positive and greater in value than $\Delta H/T$ for ΔS_{total} to be positive and the reaction to be thermodynamically feasible.
- ★ Exothermic reactions: ΔH is negative and hence $\Delta S_{surroundings}$ is positive. If ΔS_{system} is positive, the reaction is always feasible. If ΔS_{system} is negative, the reaction is only feasible if the positive $\Delta S_{suroundings}$ outweighs the negative ΔS_{system}.
- ★ The magnitude of $\Delta S_{surroundings}$ decreases as the temperature rises and so its contribution to ΔS_{total} gets less.

Hint

Remember: $\Delta S_{surroundings} = -\Delta H/T$

Worked example

Consider the reaction:

$$CH_4(g) + H_2O(g) \longrightarrow CO(g) + 3H_2(g) \qquad \Delta H = +206\,kJ\,mol^{-1}$$

a Is ΔS_{system} likely to be positive or negative?

b Use the data below to calculate the feasibility of this reaction at 750°C.

c Will the reaction be more or less feasible at a lower temperature?

	$CH_4(g)$	$H_2O(g)$	$CO(g)$	$H_2(g)$
$S/J\,K^{-1}\,mol^{-1}$	186	189	198	131

Answer

a 2 mol of gas produce 4 mol of gas, so disorder increases. Therefore, ΔS_{system} will be positive.

b $\Delta S_{system} = 3 \times 131 + 198 - (186 + 189) = +216\,J\,K^{-1}\,mol^{-1}$

$\Delta S_{surroundings} = -\Delta H/T = -(+206\,000/1023) = -201\,J\,K^{-1}\,mol^{-1}$

$\Delta S_{total} = \Delta S_{system} + \Delta S_{surroundings} = +216 + (-201) = +15\,J\,K^{-1}\,mol^{-1}$

ΔS_{total} is positive and so the reaction is feasible at 750°C.

c At a lower temperature, $\Delta S_{surroundings}$ will be more negative and so the reaction will be less feasible.

Hint

Do not forget to change 206 kJ to 206 000 J and 750°C to 1023 K.

Solubility of ionic compounds

Calculation of $\Delta H_{solution}$

- ★ This is a balance between the lattice energy and the sum of the enthalpies of hydration of the ions:
  $\Delta H_{solution} = -\text{lattice energy} + \Delta H_{hydration}(\text{cation}) + \Delta H_{hydration}(\text{anion})$
- ★ This is shown by the Hess's law diagram in Figure 4.1.

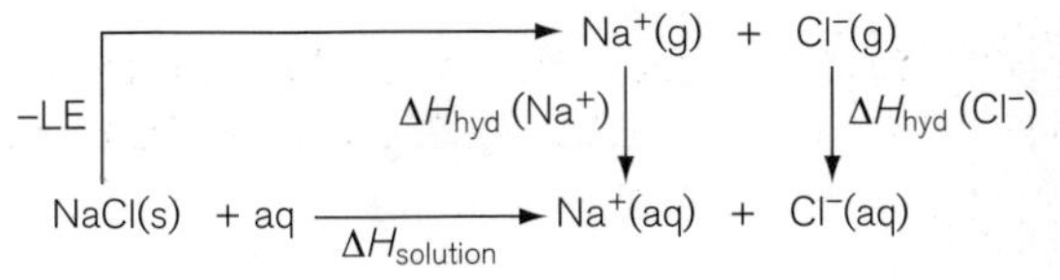

$Na^+(g) + Cl^-(g)$

$-LE$ $\quad \Delta H_{hyd}(Na^+) \quad \Delta H_{hyd}(Cl^-)$

$NaCl(s) + aq \xrightarrow{\Delta H_{solution}} Na^+(aq) + Cl^-(aq)$

Figure 4.1 Hess's law diagram for dissolving sodium chloride

Factors affecting lattice energy

- ★ Charge on the ions: the greater the charge, the more exothermic the lattice energy.
- ★ The sum of the ionic radii of the ions: the smaller the radii, the more exothermic the lattice energy.

Factors affecting enthalpy of hydration

- ★ Charge on the ion: the greater the charge, the more exothermic the enthalpy of hydration.
- ★ Radius of the ion: the smaller the radius, the more exothermic the enthalpy of hydration.

The reason that anhydrous group 2 compounds have a negative ΔS_{system} is that the ions become fully hydrated, for example:

$MgSO_4(s) + 6H_2O(l) \longrightarrow [Mg(H_2O)_6]^{2+}(aq) + SO_4^{2-}(aq)$

Seven particles become two, so entropy decreases.

ΔS_{total} and hence solubility

- ★ ΔS_{system} can be worked out in the usual way. It is positive for group 1 compounds and negative for anhydrous group 2 compounds.
- ★ $\Delta S_{surroundings} = -\Delta H/T$ and so $\Delta S_{total} = \Delta S_{system} - \Delta H/T$
- ★ If ΔS_{total} is positive, the solid will dissolve; if negative, it will not.

Checklist

Before attempting the questions on this topic, check that you:

- ☐ know that entropy increases in the order solid ⟶ liquid ⟶ gas and as temperature is increased
- ☐ can calculate ΔS_{system}, $\Delta S_{surroundings}$ and hence ΔS_{total}
- ☐ understand that ΔS_{total} must be positive for a reaction to be feasible
- ☐ can explain the effect of a change in temperature on the value of $\Delta S_{surroundings}$ and hence on the feasibility of the reaction
- ☐ know the definitions of lattice energy and enthalpy of hydration and the factors that affect their value
- ☐ can calculate $\Delta H_{solution}$ from lattice energy and hydration enthalpy data

Testing your knowledge and understanding

For the following set of questions, cover the margin, write your answers, then check to see if you are correct.

- ★ Which has the highest entropy: $H_2O(l)$ at 25°C, $H_2O(g)$ at 25°C, $H_2O(l)$ at 100°C or $H_2O(g)$ at 100°C?
- ★ Explain why is there a decrease in entropy when magnesium burns in oxygen:
  $2Mg(s) + O_2(g) \longrightarrow 2MgO(s)$
- ★ Explain why is there an increase in entropy in the reaction:
  $CuSO_4.5H_2O(s) \longrightarrow CuSO_4(s) + 5H_2O(l)$

Answers

$H_2O(g)$ at 100°C

Because a random (high entropy) gas is turning into an ordered (low entropy) solid

Because one particle becomes six particles and because a solid is turning into a liquid (and another solid).

The **answers** to the **numbered questions** are on page 139

1 Methane reacts with chlorine under certain conditions as shown by the equation:
$CH_4(g) + 2Cl_2 \rightarrow CH_2Cl_2(g) + 2HCl(g)$

Values at 25°C	$CH_4(g)$	$Cl_2(g)$	$CH_2Cl_2(g)$	HCl(g)
S/J K^{-1} mol^{-1}	186	165	178	187
ΔH_f/kJ mol^{-1}	−75	0	−124	−92

Calculate:
a ΔS_{system}
b ΔH and hence $\Delta S_{surroundings}$ at 25°C
c ΔS_{total} and hence comment on the feasibility of this reaction

2 Calculate the enthalpy of solution of calcium chloride, $CaCl_2(s)$.
Data: lattice energy ($CaCl_2$) = −2237 kJ mol^{-1}
enthalpies of hydration of Ca^{2+} = −1650 and of Cl^- = −364 kJ mol^{-1}

3 Calculate whether silver fluoride, AgF and silver chloride, AgCl, are soluble in water at 25°C.

	ΔS_{system}/J K^{-1} mol^{-1}	$\Delta H_{solution}$/kJ mol^{-1}
AgF(s)	−21	−20
AgCl(s)	+34	+66

4 Explain:
a why the lattice energy of calcium sulfate, $CaSO_4$, is more exothermic than that of barium sulfate, $BaSO_4$
b why the enthalpy of hydration of the magnesium ion is more exothermic than that of the sodium ion

Topic 4.3 Equilibria and application of rates and equilibrium

Introduction

- ★ Concentration, in the context of equilibrium, is measured in mol dm^{-3} and the concentration of a substance A is written as [A].
- ★ The common errors in this topic are to use moles, rather than concentrations, and initial values, rather than equilibrium values, when substituting in expressions for K_c.
- ★ Calculations of K must be set out clearly, showing each step.

Things to learn

- ★ A homogeneous equilibrium is one in which all the substances are in the same phase. In a heterogeneous equilibrium there are two or more phases (usually solid and gas).
- ★ An increase in temperature causes $\Delta S_{surroundings}$ to become less positive for an exothermic reaction and less negative for an endothermic reaction.
- ★ The value of K for a reaction is altered only by a change in temperature.

Things to understand

Dynamic equilibrium

★ At equilibrium the rate of the forward reaction equals the rate of the back reaction, so there is no change in the concentrations of any of the species.

K_c, the equilibrium constant measured in terms of concentrations

K_c is only equal to the quotient
$$\frac{[C]^x[D]^y}{[A]^m[B]^n}$$
when the system is at equilibrium. If the quotient does *not* equal K_c, the system is *not* in equilibrium and the reaction will continue until equilibrium is reached, when there will be no further change in any of the concentrations.

★ K_c is found from the chemical equation.
For a reaction: $mA + nB \rightleftharpoons xC + yD$
where m, n, x and y are the stoichiometric numbers in the equation:
$$K_c = \frac{[C]^x_{eq}[D]^y_{eq}}{[A]^m_{eq}[B]^n_{eq}}$$
All the concentrations are equilibrium values.

Calculation of K_c

The calculation is carried out in five steps:

(1) Draw up a table and fill in the initial number of moles, the change for each substance and the equilibrium number of moles of each substance.
(2) Convert equilibrium moles to concentration in mol dm^{-3}.
(3) State the expression for K_c.
(4) Substitute equilibrium concentrations into the expression for K_c.
(5) Work out the units for K_c and add them to your answer.

Worked example

An important reaction in the blast furnace is the formation of carbon monoxide:

$$C(s) + CO_2(g) \rightleftharpoons 2CO(g)$$

This is an example of a heterogeneous equilibrium.

When 1.0 mol of carbon dioxide was heated with excess carbon at a temperature of 700°C in a vessel of volume 20 dm^3, 95% of the carbon dioxide reacted. Calculate K_c.

Answer

	CO_2(g)	CO(g)	units
Initial amount	1.0	0	mol
Change	−0.95	+0.95 × 2	mol
Equilibrium amount	1.0 − 0.95 = 0.05	1.9	mol
Equilibrium concentration	0.05/20 = 0.0025	1.9/20 = 0.095	mol dm^{-3}

$$K_c = \frac{[CO]^2_{eq}}{[CO_2]_{eq}} = \frac{(0.095)^2(\text{mol dm}^{-3})^2}{0.0025\,\text{mol dm}^{-3}} = 3.6\,\text{mol dm}^{-3}$$

C(s) does not appear in the expression for K_c because it is a solid. The amount of CO produced is twice the amount of CO_2 reacted. The units can be worked out by dimensions. In this example, they are $(\text{concentration})^2$ divided by (concentration) and so are mol dm^{-3}.

Partial pressure

★ The partial pressure of a gas A, p(A), in a mixture is the pressure that the gas would exert if it alone filled the container. It is calculated from the expression:

partial pressure of a gas = mole fraction of that gas × total pressure

where

$$\text{mole fraction} = \frac{\text{number of moles of that gas}}{\text{total number of moles of gas}}$$

- ★ The total pressure, P, is equal to the sum of the partial pressures of each gas in the mixture.

K_p, the equilibrium constant measured in terms of partial pressure

- ★ This only applies to reactions involving gases. Solids and liquids do *not* appear in the expression for K_p.
  For a reaction: $mA(g) \rightleftharpoons xB(g) + yC(g)$
  where m, x and y are the stoichiometric numbers in the equation:
  $$K_p = \frac{p(B)^x_{eq}\, p(C)^y_{eq}}{p(A)^m_{eq}}$$
  So for the reaction: $CH_4(g) + H_2O(g) \rightleftharpoons CO(g) + 3H_2(g)$
  $$K_p = \frac{p(CO)_{eq}\, p(H_2)^3_{eq}}{p(CH_4)_{eq}\, p(H_2O)_{eq}} \qquad \text{units} = \frac{\text{atm}^3}{\text{atm}^2} = \text{atm}$$

Calculations involving K_p

The calculation of K_p from equilibrium data is carried out in a similar way to that for K_c.

(1) Draw up a table and fill in the initial number of moles, the change for each substance, the equilibrium number of moles of each substance and the total number of moles at equilibrium.
(2) Convert equilibrium moles to mole fraction.
(3) Multiply the mole fraction of each by the **total** pressure.
(4) State the expression for K_p.
(5) Substitute partial pressures into the expression for K_p.
(6) Work out the units for K_p and add them to your answer.

Worked example

0.080 mol of PCl_5 was placed in a vessel and heated to 275°C. When equilibrium had been reached, it was found that the total pressure was 2.0 atm and that 40% of the PCl_5 had dissociated.

Calculate K_p for the reaction: $PCl_5(g) \rightleftharpoons PCl_3(g) + Cl_2(g)$

Answer

	PCl_5	PCl_3	Cl_2	Total
Initial amount/mol	0.080	0	0	
Change/mol	– 0.032	+ 0.032	+ 0.032	
Equilibrium amount/mol	0.080 x 0.032 = 0.048	0.032	0.032	0.112
Mole fraction	0.048/0.112 = 0.429	0.032/0.112 = 0.286	0.032/0.112 = 0.286	
Partial pressure/atm	0.429 x 2 = 0.857	0.286 x 2 = 0.571	0.286 x 2 = 0.571	

As 40% of the PCl_5 had dissociated, 0.40 × 0.080 = 0.032 mol had reacted and 0.032 mol of PCl_3 and Cl_2 were produced.

$$K_c = \frac{p(PCl_3)_{eq}\, p(Cl_2)_{eq}}{p(PCl_5)_{eq}} = \frac{0.571\ \text{atm} \times 0.571\ \text{atm}}{0.857\ \text{atm}} = 0.38\ \text{atm}$$

Hint

If no initial amount of PCl_5 is given, you should call it 1 mol and then do the calculation in the same way.

Relationship between entropy and equilibrium constant

R is the gas constant and equals 8.31 J K^{-1} mol^{-1}.

$\Delta S_{total} = R \ln K$, so $K = e^{\Delta S_{total}/R}$. A positive ΔS gives a value of $K > 1$ and a negative ΔS gives a value of $K < 1$.

Extent of reaction

The more positive is ΔS_{total} the larger is the value of K, and so the equilibrium lies more to the right.

ΔS_{total}/J K^{-1} mol^{-1}	Equilibrium constant, K	Extent of reaction
Greater than +150	Greater than 10^8	Almost complete: almost no reactants remain
Between +60 and +150	Between 10^8 and 10^3	Reaction greatly favours products
Between +60 and −60	Between 10^3 and 10^{-3}	Significant amounts of both reactants and products
Between −60 and −150	Between 10^{-3} and 10^{-8}	Reaction greatly favours reactants
Less than −150	Less than 10^{-8}	Reaction not noticeable

Application of rates and equilibrium

Variation of *K* and yield with conditions

Temperature

This is the *only* factor that alters the value of K.

If ΔS_{total} becomes smaller, $\ln K$ and hence K will also become smaller because $\Delta S_{total} = R \ln K$.

- If a reaction is exothermic, an increase in temperature makes $\Delta S_{surroundings}$ less positive. This will make ΔS_{total} smaller and hence lower the value of K. The result is that the quotient (concentration or partial pressure expression) is now larger than the new K. The system reacts to reduce the quotient by moving the position of equilibrium to the left (the endothermic direction).

K only equals the quotient when the system is at equilibrium. If the quotient is greater than K, the reaction will move to the left until the two are equal. If the quotient is smaller than K, the reaction will move to the right until the two are equal.

- If a reaction is endothermic left to right, an increase in temperature makes $\Delta S_{surroundings}$ less negative. This will make ΔS_{total} larger and hence increase the value of K. The result is that the quotient (concentration or partial pressure expression) is now smaller than the new K. The system reacts to increase the quotient by moving the position of equilibrium to the right (the endothermic direction), increasing the yield.
- An increase in temperature will increase the rates of both the forward and back reactions, so equilibrium is reached more rapidly.

Pressure

- A change in pressure does *not* alter K. If there are more gas molecules on one side of the equation than the other, the value of the quotient will be altered by a change in pressure. Therefore, the reaction will no longer be at equilibrium and will continue until the value of the quotient once again equals K. If there are more gas molecules on the left-hand side of the equation and the pressure is increased, the quotient will get smaller and the position of equilibrium will move to the right until the quotient equals the unchanged value of K.

Concentration

- A change in the concentration of one of the substances in the equilibrium mixture does *not* alter the value of K, but it *does* alter the value of the quotient. Therefore, the reaction will no longer be at equilibrium. It will continue until the value of the quotient once again equals K. If the concentration of a reactant on the left-hand side of the equation is increased, the position of equilibrium will move to the right.

Catalyst

- This neither alters the value of K nor the quotient and so has no effect on the position of equilibrium. It speeds up the forward and the reverse reactions equally. Thus it causes equilibrium to be reached more quickly.

Checklist

Before attempting questions on this topic, check that you:

- ☐ can explain the meaning of dynamic equilibrium
- ☐ can deduce the expressions for K_c and K_p and its units, given the equation
- ☐ can calculate the values of K_c and K_p given equilibrium data
- ☐ can calculate the value of K given entropy and enthalpy data
- ☐ know not to include values for solids and liquids in the expressions for K_c and for K_p in gaseous systems
- ☐ know that *only* temperature can alter the value of K, and how it affects the position of equilibrium

Testing your knowledge and understanding

For the following questions, cover the margin, write your answers, then check to see if you are correct.

- Dry air at a pressure of $104\,kN\,m^{-2}$ contains 78.1% nitrogen 21.0% oxygen and 0.9% argon by moles. Calculate the partial pressures of each gas.

- Write the expression for K_c, stating its units, for the following reactions:
  - **a** $2SO_2(g) + O_2(g) \rightleftharpoons 2SO_3(g)$
  - **b** $2HCl(g) + ½O_2(g) \rightleftharpoons H_2O(g) + Cl_2(g)$

- Write the expression for K_p, stating its units, for the following reactions:
  - **a** $CO(g) + 2H_2(g) \rightleftharpoons CH_3OH(g)$
  - **b** $Fe_2O_3(s) + CO(g) \rightleftharpoons 2FeO(s) + CO_2(g)$

1 A 0.0200 mol sample of sulfur trioxide was introduced into a vessel of volume $1.52\,dm^3$ at 1000°C. After equilibrium had been reached, 0.0142 mol of sulfur trioxide was found to be present. Calculate the value of K_c for the reaction:

$2SO_3(g) \rightleftharpoons 2SO_2(g) + O_2(g)$.

2 1.00 mol of nitrogen(II) oxide, NO, and 1.00 mol of oxygen were mixed in a container and heated to 450°C. At equilibrium, the number of moles of oxygen was found to be 0.70 mol. The total pressure in the vessel was 4.0 atm. Calculate the value of K_p for the reaction:

$2NO(g) + O_2(g) \rightleftharpoons 2NO_2(g)$

3 This question concerns the equilibrium reaction:

$2SO_2(g) + O_2(g) \rightleftharpoons 2SO_3(g) \qquad \Delta H = -196\,kJ\,mol^{-1}$

$K_c = 3 \times 10^4\,mol^{-1}\,dm^3$ at 450°C

Answers

$p(N_2) = 0.781 \times 104 = 81.2\,kN\,m^{-2}$

$p(O_2) = 0.210 \times 104 = 21.8\,kN\,m^{-2}$

$p(Ar) = 0.009 \times 104 = 0.9\,kN\,m^{-2}$

a $K_c = \frac{[SO_3]^2}{[SO_2]^2[O_2]}$ units: $mol^{-1}\,dm^3$

b $K_c = \frac{[H_2O][Cl_2]}{[HCl]^2[O_2]^{1/2}}$

units: $mol^{-1/2}\,dm^{3/2}$

a $K_p = \frac{p(CH_3OH)}{p(CO)(pH_2)^2}$ units: atm^{-2}

b $K_p = \frac{p(CO_2)}{p(CO)}$ no units

The **answers** to the **numbered questions** are on page 139

a 2 mol of sulfur dioxide, 1 mol of oxygen and 2 mol of sulfur trioxide were mixed in a vessel of volume 10 dm^3 at 450°C in the presence of a catalyst. State whether or not these substances are initially in equilibrium. If not, explain how the system would react.

b State and explain the effects of the following on the value of K_c and hence on the amount of sulfur dioxide converted:

(i) decreasing the temperature

(ii) decreasing the pressure

(iii) adding more catalyst

(iv) removing sulfur trioxide

4 Consider the equilibrium reaction:

$PCl_5(g) \rightleftharpoons PCl_3(g) + Cl_2(g)$ $\Delta H = +124\ kJ\ mol^{-1}$

$\Delta S_{system} = +215\ J\ K^{-1}\ mol^{-1}$

$R = 8.31\ J\ K^{-1}\ mol^{-1}$

a Calculate the value of ΔS_{total} and hence K at 25°C and at 400°C.

b Comment on the extent of the reaction at these two temperatures.

Topic 4.4 Acid–base equilibria

Introduction

★ It is essential that you can write the expression for K_a of a weak acid.

★ Make sure that you know how to use your calculator to evaluate logarithms and turn pH and pK_a values into $[H^+]$ and K_a values respectively.

★ Always give pH values to 2 decimal places.

★ Buffer solutions do *not* have a constant pH. They resist changes in pH.

★ You may use H^+, $H^+(aq)$ or H_3O^+ as the formula of the hydrogen ion.

Things to learn

★ A **Brønsted–Lowry acid** is a substance that donates an H^+ ion (a proton) to another species.

★ A **Brønsted–Lowry base** is a substance that accepts an H^+ ion from another species.

★ $\boldsymbol{K_w} = [H^+][OH^-] = 1.0 \times 10^{-14}\ mol^2\ dm^{-6}$ at 25°C

★ **A neutral** solution is one in which $[H^+] = [OH^-]$

★ **pH** $= -\log_{10}[H^+]$; **pOH** $= -\log_{10}[OH^-]$; $pK_a = -\log K_a$

★ **pH + pOH = 14** at 25°C

★ **A strong acid** is *totally* ionised in aqueous solution; a **weak acid** is only *slightly* ionised in aqueous solution.

★ For a weak acid, HA, $\boldsymbol{K_a} = \frac{[H^+][A^-]}{[HA]}$ or $\frac{[H_3O^+][A^-]}{[HA]}$

★ **A buffer** solution is a solution of known pH that has the ability to resist a change in pH when small amounts of acid or base are added.

Things to understand

Conjugate acid–base pairs

★ These are linked by an H^+ ion:

acid minus $H^+ \rightarrow$ conjugate base. For example:

CH_3COOH (acid) $- H^+ \rightarrow CH_3COO^-$ (conjugate base)

base + H^+ → its conjugate acid — for example:

NH_3 (base) + H^+ → NH_4^+ (conjugate acid)

- When sulfuric acid is added to water, it acts as an acid and the water acts as a base:

  $H_2SO_4 + H_2O \rightarrow H_3O^+ + HSO_4^-$

  acid + base → conjugate acid of H_2O + conjugate base of H_2SO_4
- Hydrogen chloride gas acts as an acid when added to ammonia gas:

  $HCl + NH_3 \rightarrow NH_4^+ + Cl^-$

  acid + base → conjugate acid of NH_3 + conjugate base of HCl

pH scale

- Water partially ionises:

  $H_2O(l) \rightleftharpoons H^+(aq) + OH^-(aq)$

  The $[H_2O]$ is very large and is, therefore, effectively constant, so its value can be incorporated into K for the reaction:

  $K_w = [H^+][OH^-] = 1.0 \times 10^{-14}\ mol^2\ dm^{-6}$ at 25°C

If the acidity of a solution increases, $[H^+]$ increases and pH decreases.

- A neutral solution is one in which $[H^+] = [OH^-]$; pH = 7 at 25°C
- An acidic solution is one in which $[H^+] > [OH^-]$; pH < 7 at 25°C
- An alkaline solution is one in which $[H^+] < [OH^-]$; pH > 7 at 25°C

pH of strong acids and bases

The second ionisation of a dibasic acid such as H_2SO_4 is weak. So $[H^+]$ for 0.321 mol dm^{-3} H_2SO_4 is *not* 0.642 mol dm^{-3} — it is only slightly greater than 0.321 mol dm^{-3}.

- For a strong monobasic acid:

  $pH = -\log_{10}[acid]$

  Therefore, the pH of a 0.321 mol dm^{-3} solution of HCl = –log 0.321 = 0.49
- For a strong alkali with one OH^- in its formula:

  $pOH = -\log_{10}[alkali]$ and pH = 14 – pOH

  Therefore, the pH of a 0.109 mol dm^{-3} solution of NaOH is calculated as follows:

  pOH = –log 0.109 = 0.96

  pH = 14 – 0.96 = 13.04

For an alkali with two OH^- per formula, the $[OH^-]$ is 2 × [alkali]. The pH of a 0.109 mol dm^{-3} solution of $Ba(OH)_2$ is calculated as follows:
pOH = –log (2 × 0.109) = 0.66
pH = 14 – 0.66 = 13.34

Hint

Remember that acids have pH < 7 and alkalis have pH > 7 at 25°C.

pH of weak acids

- Weak acids ionise partially. The general equation is:

  $HA(aq) \rightleftharpoons H^+(aq) + A^-(aq)$
- The acid produces *equal* amounts of H^+ and A^-, so $[H^+] = [A^-]$
- This enables either the pH of a weak acid solution or the value of K_a for a weak acid to be calculated.

This assumes that there are no other sources of A^- or H^+. This is *not* true in a buffer solution.

A second assumption is that the value of [HA] is the same as its initial concentration. This is a fair assumption, as long as less than 5% of the acid ionises.

Worked example

Calculate the pH of a 0.123 mol dm^{-3} solution of an acid, HA, that has $K_a = 4.56 \times 10^{-5}$ mol dm^{-3}.

Answer

$HA(aq) \rightleftharpoons H^+(aq) + A^-(aq)$

$$K_a = \frac{[H^+][A^-]}{[HA]} = \frac{[H^+]^2}{[HA]}$$

$$[H^+] = \sqrt{K_a[HA]}$$

$$= \sqrt{4.56 \times 10^{-5} \times 0.123} = 2.37 \times 10^{-3}\ mol\ dm^{-3}$$

$pH = -\log(2.37 \times 10^{-3}) = 2.63$

Effect of dilution

- ★ Strong acids: if diluted by a factor of 10, the pH increases by 1 unit.
- ★ Weak acids: if diluted by a factor of 10, the pH increases by 0.5 units.
- ★ Buffer solutions: if diluted, the pH does *not* change.

Titration curves

- ★ These are drawn either for the addition of base to a volume $V\,cm^3$ of a 0.1 mol dm^{-3} solution of acid until the base is in excess, or for addition of acid to $V\,cm^3$ of a 0.1 mol dm^{-3} solution base (see Figure 4.2).
- ★ If the acid and base are of the same concentration, the end point is at a volume, $V\,cm^3$.
- ★ There are several pH values to remember. These are shown in the table below.

Titration	Starting pH	End point pH	Vertical pH range	Final pH
Strong acid by strong base	1	7	3.5–10.5	13
Weak acid by strong base	3	≈ 9	7–10.5	13
Strong acid by weak base	1	≈ 5	3.5–7	11

At halfway to the end point ($12.5\,cm^3$ added if the end point is at $25\,cm^3$), the curve for a weak acid–strong base titration is almost horizontal. This is when [HA] = $[A^-]$. Thus the halfway pH = pK_a for the weak acid.

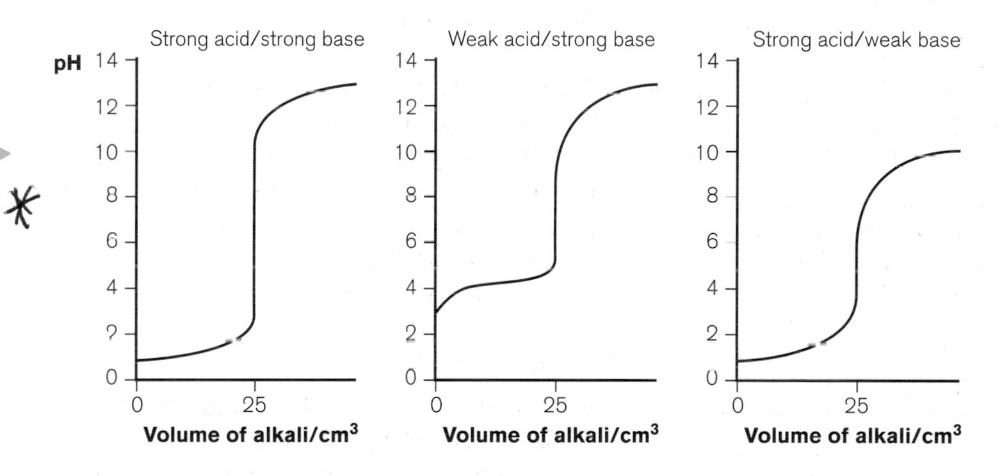

Figure 4.2 Titration curves

Indicators

- ★ To be of use in a titration, the colour of the indicator must change completely within the pH of the *vertical* part of the titration curve.

Indicators	Vertical range of pH	Suitable indicator
Strong acid–strong base	3.5–10.5	Methyl orange or phenolphthalein
Weak acid–strong base	7–10.5	Phenolphthalein
Strong acid–weak base	3.5–7	Methyl orange

Buffer solutions

- ★ These consist of an acid–base conjugate pair, e.g. a weak acid and its salt, such as CH_3COOH and CH_3COONa, or a weak base and its salt, such as NH_3 and NH_4Cl.

- To be able to resist pH changes, the concentration of the acid and its conjugate base must be similar and large relative to the small amounts of H^+ or OH^- being added.
- Consider a buffer of ethanoic acid and sodium ethanoate.

  The salt is fully ionised:

  $CH_3COONa(aq) \longrightarrow CH_3COO^-(aq) + Na^+(aq)$

  The weak acid is slightly ionised:

  $CH_3COOH(aq) \rightleftharpoons H^+(aq) + CH_3COO^-(aq)$

  If small amounts of H^+ are added to the solution, almost all of them are removed by reaction with the large reservoir of CH_3COO^- ions:

  $H^+(aq) + CH_3COO^-(aq) \longrightarrow CH_3COOH(aq)$

  $[CH_3COOH]$ and $[CH_3COO^-]$ change by insignificant amounts so their ratio and hence the pH of the solution hardly alters.

  If some OH^- ions are added, almost all of them are removed by reaction with the weakly acidic CH_3COOH molecules:

  $OH^-(aq) + CH_3COOH \longrightarrow CH_3COO^-(aq) + H_2O(l)$

  $[CH_3COOH]$ and $[CH_3COO^-]$ change by insignificant amounts so their ratio and hence the pH of the solution hardly alters.

Calculation of the pH of a buffer solution

Consider a buffer solution made from a weak acid, HA, and its salt NaA.

The acid is slightly ionised:

$HA(aq) \rightleftharpoons H^+(aq) + A^-(aq)$

Thus, $K_a = \dfrac{[H^+][A^-]}{[HA]}$

The salt is totally ionised, therefore:

$[A^-(aq)] = [salt]$ and $[HA(aq)] = [weak\ acid]$

$K_a = \dfrac{[H^+][salt]}{[weak\ acid]}$ or $[H^+] = K_a \dfrac{[weak\ acid]}{[salt]}$

$pH = -\log[H^+]$

> If the concentrations of the weak acid and its salt are the same, then $[H^+] = K_a$ and $pH = pK_a$.

Worked example

Calculate the pH of 500 cm^3 of a solution containing 0.121 mol of ethanoic acid and 0.100 mol of sodium ethanoate.

pK_a for ethanoic acid = 4.76

Answer

$pK_a = 4.76$

$K_a = 10^{-pKa} = 1.74 \times 10^{-5}\ mol\ dm^{-3}$

[weak acid] = 0.121/0.500 = 0.242 mol dm^{-3}

[salt] = 0.100/0.500 = 0.200 mol dm^{-3}

$[H^+] = \dfrac{K_a[weak\ acid]}{[salt]} = \dfrac{1.74 \times 10^{-5} \times 0.242}{0.200} = 2.10 \times 10^{-5}\ mol\ dm^{-3}$

$pH = -\log[H^+] = 4.68$

Checklist

Before attempting the questions on this topic, check that you

- ☐ can identify acid–base conjugate pairs
- ☐ can define pH and K_w

- [ ] can define K_a and pK_a for weak acids
- [ ] understand what is meant by the terms 'strong' and 'weak' as applied to acids and bases
- [ ] can calculate the pH of solutions of strong acids, strong bases and weak acids
- [ ] can recall the titration curves for the neutralisation of strong and weak acids
- [ ] can use the titration curve to calculate the value of K_a for a weak acid
- [ ] understand the reasons for the choice of indicator in acid–base titrations
- [ ] can define a buffer solution, explain its mode of action and calculate its pH

Testing your knowledge and understanding

For the following questions, cover the margin, write your answers, then check to see if you are correct.

★ Identify the acid–base conjugate pairs in the reaction:

$H_2SO_4 + CH_3COOH \rightarrow CH_3COOH_2^+ + HSO_4^-$

Answers

Acid H_2SO_4: its conjugate base HSO_4^-

Base CH_3COOH: its conjugate acid $CH_3COOH_2^+$

★ Calculate the pH of the following solutions:

a 0.11 mol dm^{-3} HCl

b 0.11 mol dm^{-3} LiOH

c 0.11 mol dm^{-3} $Ba(OH)_2$

a 0.96

b 13.04

c 13.34

The **answers** to the **numbered questions** are on page 140

1 Calculate the pH of 0.22 mol dm^{-3} C_2H_5COOH, which has a pK_a value of 4.87.

2 25 cm^3 of a weak acid, HX, of concentration 0.10 mol dm^{-3} was titrated with 0.10 mol dm^{-3} sodium hydroxide solution and the pH was measured at intervals. The results are shown in the table below.

Volume NaOH/cm^3	5	10	12	20	23	24	25	26	30
pH	4.5	4.8	4.9	5.5	6.5	7.0	9.0	12.0	12.5

a Draw the titration curve and use it to calculate pK_a for the acid, HX.

b Suggest a suitable indicator for the titration.

3 a Define 'buffer solution'. Give the names of a pair of substances that act as a buffer when in solution.

b Explain how a buffer resists change in pH when small amounts of either H^+ or OH^- ions are added.

4 Calculate the pH of a solution made by adding 4.4 g of sodium ethanoate, CH_3COONa, to 100 cm^3 of a 0.44 mol dm^{-3} solution of ethanoic acid.

K_a for ethanoic acid = 1.74×10^{-5} mol dm^{-3}

Topic 4.5 Further organic chemistry

Introduction

★ You must revise structural and *E/Z* (geometric) isomerism (page 24).

★ This section deals with carbonyl compounds, acids and their derivatives.

Things to learn

★ A nucleophile is a species that has a lone pair of electrons and uses it to form a bond with a δ+ atom.

★ Stereoisomerism includes both geometric and optical isomerism.

★ One optical isomer rotates the plane of polarisation of plane-polarised light clockwise; the other isomer rotates it anti-clockwise.

Things to understand

Optical isomerism

Having four different groups attached to one carbon atom results in optical isomerism. Such a carbon atom is called a chiral centre.

★ Optical isomers are defined as isomers that are the non-superimposable mirror images of each other.

Thus 2-hydroxypropanoic acid (lactic acid), $CH_3CH(OH)COOH$, exists as two optical isomers. These must be drawn three-dimensionally to show that one is the mirror image of the other.

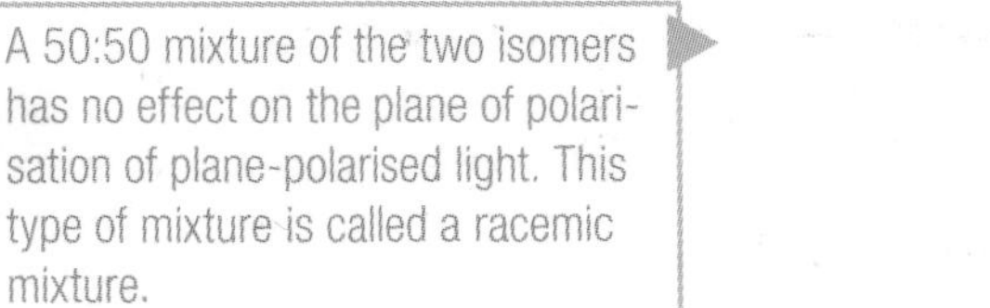
A 50:50 mixture of the two isomers has no effect on the plane of polarisation of plane-polarised light. This type of mixture is called a racemic mixture.

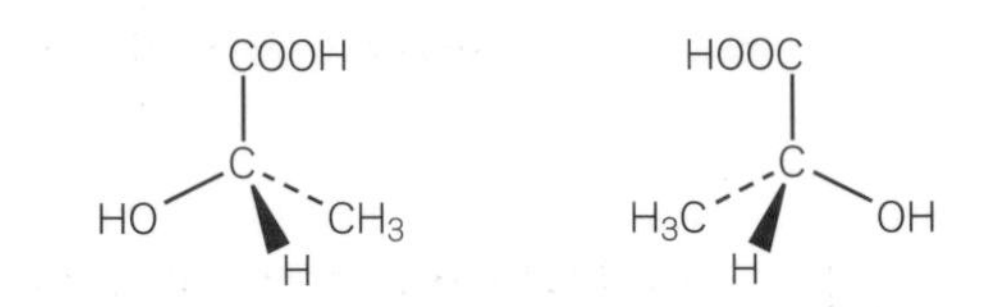

Figure 4.3 Optical isomers of 2-hydroxypropanoic acid

★ If a single optical isomer of a halogenoalkane is reacted with a nucleophile, the effect of the product on plane-polarised light depends on the mechanism (see page 58) of the reaction.

S_N1: the intermediate is planar, so the nucleophile can attack from either above or below the plane. This results in both isomers being produced in a 50:50 ratio and so the product has no effect on plane-polarised light.

S_N2: the nucleophile attacks from the side opposite to the halogen. A single chiral isomer is produced that rotates the plane of polarisation of plane-polarised light.

Carbonyl compounds (aldehydes and ketones)

These contain the >C=O functional group. Aldehydes have the general formula RCHO. Ketones the general formula RCOR′ where R and R′ are alkyl or aryl groups that may or may not be different.

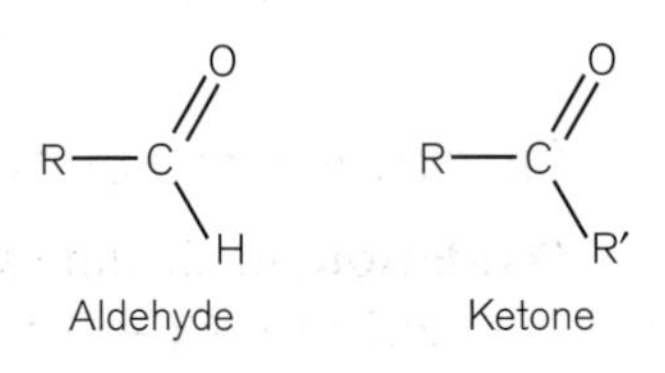

Physical properties

Neither aldehydes nor ketones can form intermolecular hydrogen bonds. Therefore, they have boiling points below those of alcohols and acids. The lower members of the homologous series are water-soluble because they can form hydrogen bonds between the lone pair of electrons on their δ– oxygen atoms and the δ+ hydrogen in water.

Reactions in common

★ **Nucleophilic addition** — for example, the addition of HCN in the presence of a CN^- ion catalyst:

$CH_3CHO + HCN \rightarrow CH_3CH(OH)CN$
2-hydroxypropanenitrile

Conditions: HCN + a trace of KOH, or HCN + KCN

> The product of this reaction of an aldehyde or an asymmetrical ketone is chiral. Since the first step is attack by CN^- on the planar carbonyl group, the product is a racemic mixture.

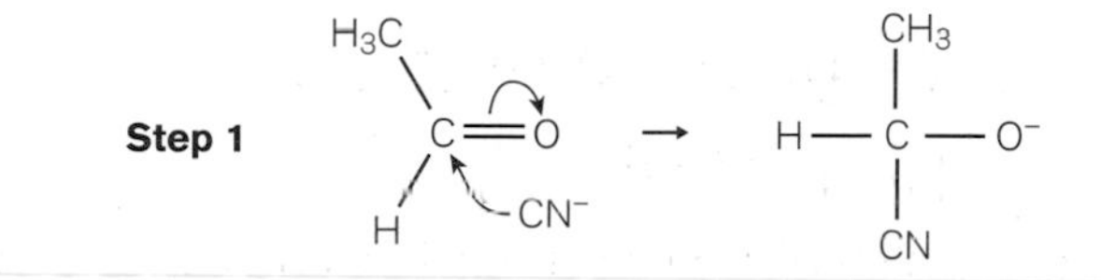

The lone pair of electrons on the carbon atom of the CN^- ion forms a bond with the δ+ carbon atom of the >C=O group.

Step 2: $H{-}C(CH_3)(CN){-}O^- \;\; H{-}CN \rightarrow H{-}C(CH_3)(CN){-}OH + CN^-$

> Aldehydes are reduced to primary alcohols and ketones to secondary alcohols.

★ **Reduction** to produce an alcohol — for example:

$CH_3CHO + 2[H] \rightarrow CH_3CH_2OH$

$CH_3COCH_3 + 2[H] \rightarrow CH_3CH(OH)CH_3$

Reagent: lithium tetrahydridoaluminate(III), dissolved in dry ether, followed by hydrolysis with $H^+(aq)$

★ **Reaction with 2,4-dinitrophenylhydrazine** (Brady's reagent) to give an orange /yellow precipitate:

> The formation of a precipitate with 2,4-DNP is a test for carbonyl compounds, i.e. for both aldehydes and ketones.

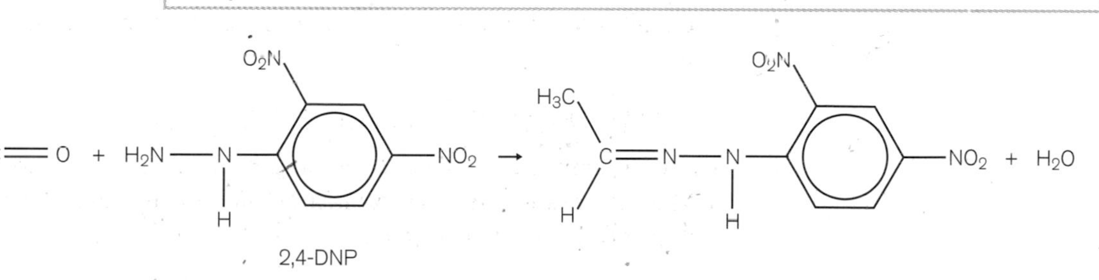

Specific reaction of aldehydes

> The formation of the silver mirror with Tollens' reagent is the test that distinguishes aldehydes, which give a positive result, from ketones, which do not react.
> Aldehydes are also oxidised by acidified potassium dichromate(VI) to form the carboxylic acid.

Oxidation, in alkaline solution, to the salt of a carboxylic acid:

$RCHO + [O] + OH^- \rightarrow RCOO^- + H_2O$

★ Tollens' reagent — on warming, the silver ions are reduced to a mirror of metallic silver

★ Fehling's (or Benedict's) solution — on warming, the blue copper(II) ions are reduced to a red precipitate of copper(I) oxide.

Iodoform reaction

★ This is a reaction that produces a pale yellow precipitate of iodoform, CHI_3, when an organic compound is added to a mixture of iodine and dilute sodium hydroxide.

> Ethanal and all ketones containing the $CH_3C{=}O$ group give a precipitate of iodoform.

- The group responsible for this reaction is the $CH_3C{=}O$ group, and the reaction involves the loss of one carbon atom from the chain.

  $C_2H_5COCH_3$ produces C_2H_5COONa and $CHI_3(s)$
- Ethanol and secondary alcohols containing the $CH_3CH(OH)$ group also undergo this reaction. Under the conditions of the reaction they are oxidised to the $CH_3C{=}O$ group, which then reacts to give iodoform.

> **Hint**
>
> This reaction is often used in organic 'problem' questions. The pale yellow precipitate produced on addition of sodium hydroxide and iodine is the clue. The organic compound that produces this precipitate will either be a carbonyl compound with a $CH_3C{=}O$ group or an alcohol with a $CH_3CH(OH)$ group. It is *not* a general test for ketones.

Carboxylic acids

These have the functional group:

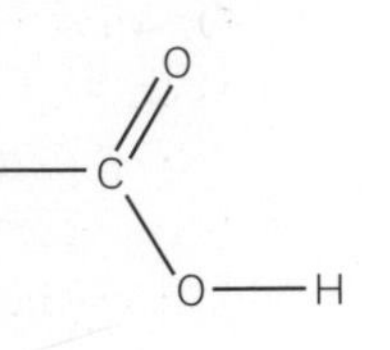

Physical properties

Carboxylic acids are liquids that have intermolecular hydrogen bonding. They can also hydrogen bond with water and so the lower members of the series are water-soluble.

Preparation

- Oxidation of a primary alcohol or aldehyde — heat under reflux with dilute sulfuric acid and excess potassium dichromate(VI).
- Hydrolysis of a nitrile (a compound with a $C{\equiv}N$ group) — heat under reflux with dilute sulfuric acid:

  $RC{\equiv}N + 2H_2O + H^+ \rightarrow RCOOH + NH_4^+$

Reactions

- **Esterification** with alcohols — for example:

  $CH_3COOH + C_2H_5OH \rightleftharpoons CH_3COOC_2H_5 + H_2O$
  ethyl ethanoate

  Conditions: heat under reflux with a few drops of concentrated sulfuric acid.
- **Reduction** to produce a primary alcohol – for example:

  $CH_3COOH + 4[H] \rightarrow CH_3CH_2OH + H_2O$
  ethanol

  Conditions: lithium tetrahydridoaluminate(III) in dry ether, followed by addition of aqueous acid.
- With **phosphorus pentachloride** to produce an acid chloride — for example:

  $CH_3COOH + PCl_5 \rightarrow CH_3COCl + HCl + POCl_3$
  ethanoyl chloride

  Conditions: dry.

  Observation: steamy fumes given off

> Carboxylic acids and alcohols give steamy fumes with PCl_5.

- **Neutralisation** — for example:

  $2CH_3COOH + Na_2CO_3 \rightarrow 2CH_3COONa + CO_2 + H_2O$

  Organic acids have the same reactions as those of acids such as sulfuric.

> The evolution of carbon dioxide gas with sodium carbonate or sodium hydrogencarbonate distinguishes an acid from an alcohol.

Acid chlorides

These have the general formula RCOCl. The functional group is:

Examples are ethanoyl chloride, CH_3COCl, and propanoyl chloride, C_2H_5COCl.

Reactions of ethanoyl chloride

- With water to produce the carboxylic acid, ethanoic acid:
  $CH_3COCl + H_2O \rightarrow CH_3COOH + HCl$
  Observation: steamy fumes of HCl

> This reaction has a higher yield, but a lower atom economy, than the reversible reaction of an acid with an alcohol.

- With alcohols to produce an ester:
  $CH_3COCl + C_2H_5OH \rightarrow CH_3COOC_2H_5 + HCl$
  Observation: characteristic smell of the ester, ethyl ethanoate.
- With ammonia to produce the amide, ethanamide:
  $CH_3COCl + 2NH_3 \rightarrow CH_3CONH_2 + NH_4Cl$
- With primary amines to produce a substituted amide:
  $CH_3COCl + C_2H_5NH_2 \rightarrow CH_3CONH(C_2H_5) + HCl$

Esters

These have the general formula RCOOR′ where R and R′ are alkyl or aryl groups, which may or may not be different.

Reactions

> The acid is a catalyst. This reaction has a low yield because it is reversible.

- **Hydrolysis**:
  a With aqueous acid to produce the organic acid and the alcohol — for example:
  $CH_3COOC_2H_5 + H_2O \rightleftharpoons CH_3COOH + C_2H_5OH$
  ethanoic acid
  Conditions: heat under reflux with dilute sulfuric acid

> This reaction has a high yield because it is *not* reversible.

  b With aqueous alkali to produce the salt of the acid and the alcohol — for example:
  $CH_3COOC_2H_5 + NaOH \rightarrow CH_3COONa + C_2H_5OH$
  sodium ethanoate
  Conditions: heat under reflux with aqueous sodium hydroxide
- **Transesterification**:
  a If an ester is mixed with an organic acid and a catalyst, an ester of the added acid is produced — for example:

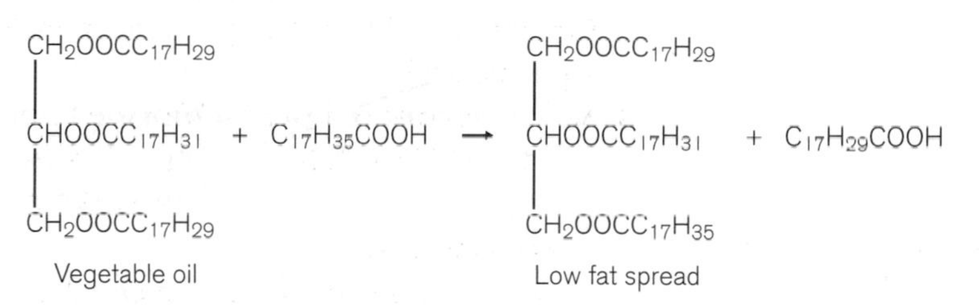

  b If mixed with a different alcohol, an ester of that alcohol is formed — for example:

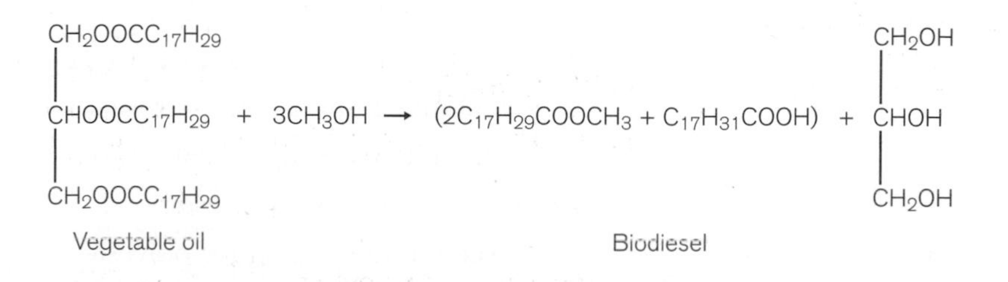

Polyesters

When an alcohol with two –OH groups is mixed with an acid with two –COOH groups or with an acid chloride with two –COCl groups, a polyester is formed in a condensation reaction. For example, with ethane-1,2-diol and benzene-1,4-dicarboxylic acid Terylene® is formed:

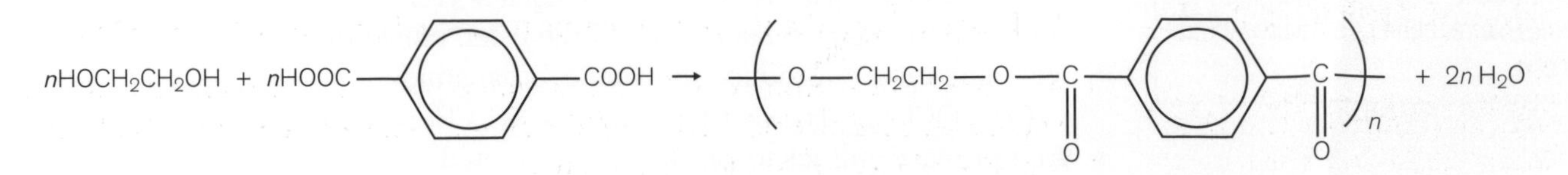

Summary of reactions

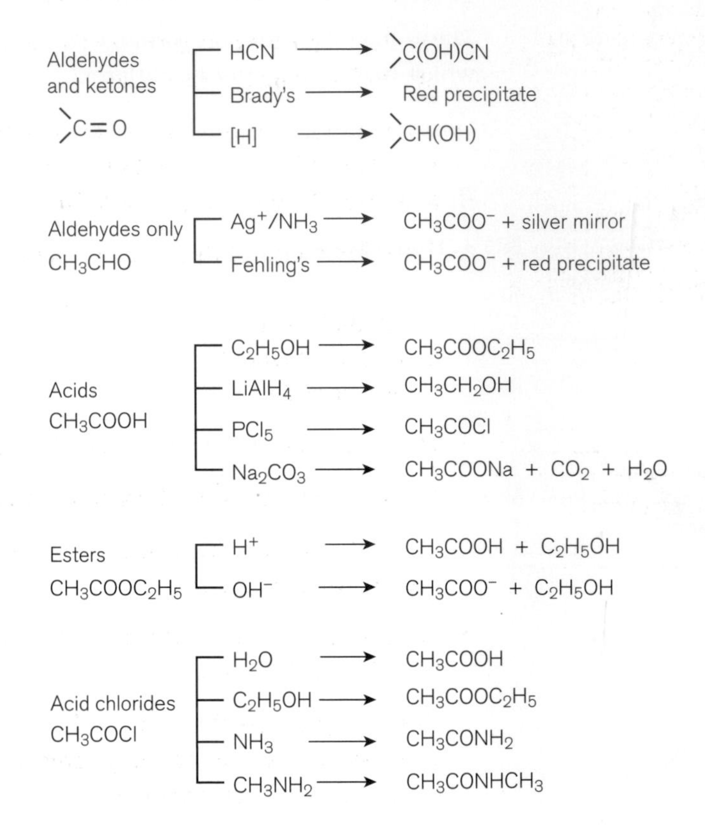

Checklist

Before attempting questions on this topic, check that you:

- ☐ can recognise stereoisomerism (geometric and optical) in organic compounds
- ☐ know the effect of an optical isomer on plane-polarised light
- ☐ understand the nature of a racemic mixture
- ☐ can recall reactions of aldehydes and ketones
- ☐ can recall the preparation and reactions of carboxylic acids
- ☐ can recall the reactions of ethanoyl chloride
- ☐ can recall the reactions of esters
- ☐ can write the repeat unit for polyesters

Testing your knowledge and understanding

For the following questions, cover the margin, write your answers, then check to see if you are correct.

- ★ State which of the following show stereoisomerism: 1,1-bromochloroethane, 2-bromopropane, 2-methylpent-2-ene, pent-2-ene
- ★ A single optical isomer of 2-chlorobutane was reacted with aqueous sodium hydroxide. The product had no effect on the plane of plane-polarised light. State the type of mechanism for this reaction.
- ★ State which of the following gives a chiral product when reacted with HCN: propanone, butan-2-one, pentan-3-one
- ★ State which of the following form intermolecular hydrogen bonds: ethanol, ethanal, ethanoic acid
- ★ Name the reagents and give the conditions for the following conversions:
  - a $C_2H_5COOCH_3$ to C_2H_5COOH and CH_3OH in a high yield
  - b C_2H_5COOH to $CH_3CH_2CH_2OH$
- ★ State the names of the reagents needed to convert ethanoyl chloride to the following:
  - a ethanoic acid
  - b methyl ethanoate
  - c ethanamide

1 Draw the stereoisomers of $CH_3CH(OH)F$.

2 Outline how you would prepare 2-hydroxypropanoic acid from ethanol.

3 Write the structural formula of the organic product of the reaction between propanone and 2,4-dintrophenylhydrazine.

4 Write the repeat unit of the polyester formed from the reaction of $HOCH_2CH_2CH_2OH$ with $ClOC(CH_2)_4COCl$.

5 Describe the tests that you could do to distinguish between the following:
- a ethanoic acid and ethanoyl chloride
- b propanal and propanone

Answers

1,1-bromochloroethane and pent-2-ene

S_N1

Butan-2-one — the others are symmetrical

Ethanol and ethanoic acid

a Heat under reflux with dilute sodium hydroxide, then add dilute sulfuric acid

b Lithium tetrahydrido-aluminate(III) in dry ether, followed by dilute sulfuric acid

a Water
b Methanol
c Ammonia

The **answers** to the **numbered questions** are on page 141

Spectroscopy and chromatography

Introduction

- ★ Spectra are used to work out the structure and identity of a substance.
- ★ Chromatography is used to separate a mixture into its constituents.

Things to learn

- ★ The frequency of an infrared line is measured by its wavenumber, with units cm^{-1}.
- ★ Absorption of infrared radiation is caused by the stretching or bending of a covalent bond that is accompanied by a change in dipole moment of the molecule.

- ★ NMR spectra are due to hydrogen nuclei in different chemical environments.
- ★ Gas chromatography (GLC) is used to separate mixtures of volatile substances.
- ★ High pressure liquid chromatography (HPLC) is used to separate mixtures of soluble materials.

Things to understand

NMR spectra

- ★ A spinning hydrogen nucleus takes up a position so that its magnetic field is either parallel to an applied magnetic field or at right angles to it. The difference in energy of these two states is equivalent to the energy of radio waves.
- ★ The number of peaks is determined by the number of different chemical environments of hydrogen atoms in the substance.
- ★ The chemical shift, δ, depends on the group in which the hydrogen atom is found.
- ★ The peaks are split according to the (n + 1) rule, where n is the number of hydrogen atoms on *adjacent* carbon atoms.
- ★ The hydrogen on an OH group is neither split nor causes splitting.

The shift due to the H in CHO is at about 9.5 ppm (parts per million) and that due to alkyl hydrogen atoms is at about 1 ppm.

Worked example

How many peaks are there in the NMR spectrum of $HOCH_2CH_2CH_2OH$ and how are they split?

Answer

There are three peaks. One is due to the CH_2 group in the middle of the molecule. This is split into five by the four hydrogen atoms on neighbouring carbon atoms. The two CH_2OH groups are in identical environments and so cause two peaks, the one due to the OH hydrogen is not split and that due to the hydrogen on the CH_2 of the CH_2OH group is split into three by the two hydrogen atoms on the middle CH_2 group.

Infrared spectra

- ★ The bending or stretching vibrations of different bonds result in absorption at different frequencies.
- ★ The C=O group in aldehydes and ketones absorbs at around 1700 cm^{-1}, and the O–H group in alcohols absorbs at over 3200 cm^{-1}.
- ★ The absorption due to O–H in alcohols and acids is broad due to hydrogen bonding.
- ★ A reaction can be followed by observing one peak disappear slowly and another peak grow steadily, e.g. the disappearance of the O–H absorption and the appearance of the C=O absorption as an alcohol is oxidised to a ketone.

See page 295 of *AS Chemistry* by George Facer, published by Philip Allan Updates, or the Edexcel data booklet for a more complete list.

Mass spectra

- ★ The peak at the highest m/e value (M) is due to the molecular ion.
- ★ The peak at m/e = M – 15 is caused by the loss of the CH_3 fragment.
- ★ The peak at m/e = M – 29 is caused by the loss of either the C_2H_5 or the CHO fragment.
- ★ The peak at m/e = M – 31 is caused by the loss of CH_2OH.

All peaks are due to positive ions and the charges must be shown on formulae.

Chromatography

This works because of the competition between a substance being adsorbed onto a stationary phase and it dissolving in a mobile phase or eluent. Thus one substance will pass through the column more quickly than another substance.

- **GLC**: the stationary phase is a liquid bonded to the inside of a coiled tube. The eluent is an unreactive gas, for example argon. The liquid mixture is injected into the top of the coiled tube which is in a thermostatically controlled oven. Different components in the mixture pass through the column one after the other.
- **HPLC**: the stationary phase is a solid packed in a straight tube. The solution mixture is injected at the top and a suitable eluent liquid is forced under high pressure through the tube. The different solutes pass through at different times.

Checklist

Before attempting questions on this topic, check that you can:

- [ ] predict the number of peaks and the splitting pattern in the NMR spectrum of a compound
- [ ] suggest how infrared spectroscopy could be used to follow a reaction
- [ ] interpret mass spectra and suggest formulae of species at different *m/e* values
- [ ] explain how GLC and HPLC work

Testing your knowledge and understanding

For the following questions, cover the margin, write your answers, then check to see if you are correct.

- How many lines are there in the NMR spectrum of pentan-2-one, $CH_3CH_2COCH_2CH_3$, and how are they split?
- What species is responsible for a peak at *m/e* = 43 in the mass spectrum of $CH_3CH_2CH_2COOH$?
- After oxidation of ethanol in alkaline solution, the mixture contained unreacted ethanol, some ethanal and some sodium ethanoate. What chromatographic method could be used to separate this mixture?

Answers

Two lines, one split into three and the other split into four.

$CH_3CH_2CH_2^+$

HPLC

1 A compound thought to be either $CH_3CH_2CH_2OH$ or $CH_2{=}CHCH_2OH$ had four lines in its NMR spectrum with a splitting pattern of a singlet, two doublets and one line split into five. Which compound is it? Justify your answer.

2 A compound of molecular formula $C_4H_{10}O$ gave steamy fumes with phosphorus pentachloride. It was oxidised with acidified potassium dichromate(VI) and the product separated and investigated. The product gave a silver mirror with Tollens' reagent and had infrared peaks at 1720 cm^{-1} and at 2900 cm^{-1} but none in the range 3650–3200 cm^{-1}. The NMR spectrum of the product had three lines. Using *all* the data, identify both the original compound and its product.

The **answers** to the **numbered questions** are on page 141

Practice Unit Test 4

Section A

1 Consider a reaction between compounds A and B. The relative rates of the reaction at different initial concentrations are shown in the table below.

Experiment	[A]	[B]	Relative rate
1	0.1	0.1	1
2	0.3	0.1	9
3	0.2	0.2	4

The rate equation is

A rate = k[A][B]
B rate = k[A][B]^2
C rate = k[A]^2[B]
D rate = k[A]^2

2 Which has the highest entropy?

A $C_2H_5OH(l)$ at 25°C
B $C_2H_5OH(l)$ at 50°C
C $CH_3CH_2CH_2OH(l)$ at 25°C
D $CH_3CH_2CH_2OH(l)$ at 50°C

3 The following reactions are all spontaneous at room temperature. Which one must be exothermic?

A $NH_3(g) + HCl(g) \rightarrow NH_4Cl(s)$
B $NH_4Cl(s) + aq \rightarrow NH_4^+(aq) + Cl^-(aq)$
C $2CH_3COOH(l) + (NH_4)_2CO_3(s) \rightarrow 2CH_3COONH_4(s) + 2H_2O(l) + CO_2(g)$
D $Ba(OH)_2.8H_2O(s) + 2NH_4Cl(s) \rightarrow BaCl_2(s) + 10H_2O(l) + 2NH_3(g)$

4 Consider the reaction:

$2Mg(s) + O_2(g) \rightarrow 2MgO(s) \qquad \Delta H = -1200\ \text{kJ mol}^{-1}$

The value of $\Delta S_{surroundings}$ at 100°C is

A more positive than at 25°C because the reaction is exothermic
B more positive than at 25°C because ΔS_{system} is negative
C less positive than at 25°C because the reaction is exothermic
D less positive than at 25°C because ΔS_{system} is negative

5 Nitrogen and hydrogen react according to the equation:

$N_2(g) + 3H_2(g) \rightleftharpoons 2NH_3(g)$

If 1.0 mol of nitrogen and 3.0 mol of hydrogen were mixed and allowed to reach equilibrium, 0.02 mol of ammonia were found to be present.

The correct number of moles of nitrogen and hydrogen at equilibrium are:

A 0.01 mol N_2 and 0.03 mol H_2
B 0.02 mol N_2 and 0.06 mol H_2
C 0.98 mol N_2 and 2.94 mol H_2
D 0.99 mol N_2 and 2.97 mol H_2

6 Consider the reaction:

$Fe_3O_4(s) + 4H_2(g) \rightleftharpoons 3Fe(s) + 4H_2O(g)$

The expression for the equilibrium constant K_p is

A $K_p = \dfrac{[Fe]^3[H_2O]^4}{[Fe_3O_4][H_2]^4}$

B $K_p = \dfrac{p(Fe)^3 p(H_2O)^4}{p(Fe_3O_4)^3 p(H_2)^4}$

C $K_p = \dfrac{p(H_2O)^4}{p(H_2)^4}$

D $K_p = \frac{[H_2O]^4}{[H_2]^4}$

7 Hydrogen is manufactured by the reaction of methane with steam:

$CH_4(g) + H_2O(g) \rightleftharpoons CO(g) + 3H_2(g) \quad \Delta H = +206\,kJ\,mol^{-1}$

The optimum conditions for this manufacture are

A low pressure, low temperature and a catalyst
B low pressure, high temperature and a catalyst
C high pressure, low temperature and a catalyst
D high pressure, high temperature and a catalyst

8 Which of the following is *not* a buffer solution when mixed together in suitable proportions?

A CH_3COOH and NaOH
B HCl and NaCl
C NH_3 and $(NH_4)_2SO_4$
D CO_2 and $NaHCO_3$

9 Which indicator would be suitable for a titration of hydrochloric acid with ammonia solution?

A thymol blue, $pK_{ind} = 1.7$
B bromocresol green, $pK_{ind} = 4.7$
C bromothymol blue, $pK_{ind} = 7.0$
D thymolphthalein, $pK_{ind} = 9.7$

10 Which of the following statements is *not* true about the solution made by adding $10\,cm^3$ of a $0.50\,mol\,dm^{-3}$ solution of sodium hydroxide to $20\,cm^3$ of a $0.50\,mol\,dm^{-3}$ solution of ethanoic acid?

A It is a buffer solution.
B The ethanoic acid and sodium hydroxide are present in a ratio of 2:1.
C Its pH is equal to the pK_a of ethanoic acid.
D Its pH remains the same on dilution

11 Which of the following does *not* exhibit stereoisomerism?

A $CH_3CH(OH)CH(OH)CH_3$
B $CH_3CH(OH)C_2H_5$
C $CH_2{=}C(CH_3)COOH$
D $CH_3CH{=}CHCOOH$

12 The product of the reaction of propanone with hydrogen cyanide has no effect on plane-polarised light. This is because

A a racemic mixture is produced
B the intermediate formed is planar
C propanone is planar at the reaction site
D the product is not chiral

13 Which if the following is *not* true when ethanal reacts with warm Fehling's solution?

A A red precipitate of Cu_2O is formed.
B The ethanal is oxidised to CH_3COOH.
C A solution containing CH_3COO^- ions is formed.
D The reaction is a redox reaction.

14 Ethanal has a lower boiling temperature than ethanoic acid because

A it cannot form intermolecular hydrogen bonds
B its covalent bonds are weaker
C it contains fewer electrons than ethanoic acid
D it is less polar than ethanoic acid

15 Which of the following isomers of $C_4H_8O_2$ will *not* form a precipitate of iodoform when warmed with iodine and alkali?

A $CH_3CH(OH)COCH_3$
B $CH_2OHCH_2COCH_3$
C $CH_2OHCH_2CH_2CHO$
D $CH_3CH(OH)CH_2CHO$

16 Ethanoyl chloride, CH_3COCl, does *not* react with

A ethanal, CH_3CHO
B ethanol, C_2H_5OH
C ethylamine, $C_2H_5NH_2$
D water, H_2O

17 The reaction of phosphorus pentachloride and CH_2OHCH_2COOH gives

A CH_2ClCH_2COOH **B** CH_2OHCH_2COCl
C CH_2ClCH_2COCl **D** CH_2ClCH_2COOCl

18 When ethyl ethanoate is warmed with

A dilute acid, an irreversible reaction forming ethanol and ethanoic acid occurs
B dilute acid, a reversible reaction forming ethanol and ethanoic acid occurs
C dilute alkali, an irreversible reaction forming ethanol and ethanoic acid occurs
D dilute alkali, a reversible reaction forming ethanol and ethanoic acid occurs

19 Which of the following *cannot* form a polyester?

A $CH_2OHCH_2CH_2OH$ and $ClOCCH_2COCl$
B $CH_2OHCH_2CH_2OH$ and $HOOCCH_2COOH$
C $ClOCCH_2COCl$ and $NH_2(CH_2)_6NH_2$
D $HOOCCH_2CH_2OH$

20 Which of the following isomers of C_4H_8O has an NMR spectrum of three lines, one a singlet, one split into three and the third split into four?

A $CH_2{=}CHCH_2CH_2OH$ **B** $CH_3CH_2COCH_3$ **C** $CH_3CH_2CH_2CHO$ **D** $(CH_3)_2CHCHO$

Section A total: 20 marks

Section B

21 a Ethanoyl chloride reacts with ethanol to form an ester, ethyl ethanoate:

$$CH_3COCl(l) + C_2H_5OH(l) \longrightarrow CH_3COOC_2H_5(l) + HCl(g) \qquad \Delta H^\circ = -21.6\,kJ\,mol^{-1}$$

(i) Calculate $\Delta S_{surroundings}$ for this reaction at 25°C. **(2 marks)**

(ii) Use the entropy data below to calculate ΔS_{system} **(1 mark)**

	$CH_3COCl(l)$	$C_2H_5OH(l)$	$CH_3COOC_2H_5(l)$	$HCl(g)$
$S^\circ/J\,K^{-1}\,mol^{-1}$	201	161	259	187

(iii) Use your answers from **(i)** and **(ii)** to calculate ΔS_{total} and hence the value of the equilibrium constant, *K*. **(2 marks)**

(iv) Comment on the feasibility and the extent of this reaction. **(2 marks)**

b Write equations for the reactions of ethanoyl chloride with the following:

(i) water **(1 mark)**

(ii) ethylamine, $C_2H_5NH_2$ **(1 mark)**

c Ethyl ethanoate can also be prepared from ethanol and ethanoic acid:

$$CH_3COOH(l) + C_2H_5OH(l) \rightleftharpoons CH_3COOC_2H_5(l) + H_2O(l) \qquad \Delta H^\circ = 0\,kJ\,mol^{-1}$$

(i) Would you expect ΔS_{system} for this reaction to be greater or smaller than that for the reaction in **a**? Justify your answer. **(1 mark)**

(ii) The number of moles of each substance present *initially* is shown in the table below.

Substance	CH_3COOH	C_2H_5OH	$CH_3COOC_2H_5$	H_2O
Moles at start	0.25	0.30	0.00	0.50

At equilibrium it was found that there were 0.12 mol of ethanoic acid present. Calculate the value of the equilibrium constant and explain why it is not necessary to know the volume of the reaction mixture. **(5 marks)**

(iii) Explain the effect, if any, of an increase in temperature on the value of the equilibrium constant, *K*, and hence on the position of equilibrium. **(2 marks)**

Total: 17 marks

22 Hydrogen cyanide, HCN, is a weak acid. Its dissociation constant, K_a, is 4.0×10^{-10} mol dm^{-3} at 25°C.

a (i) What is meant by the term 'weak' in this context? **(1 mark)**

(ii) What is the conjugate base of hydrogen cyanide? **(1 mark)**

b (i) Calculate the pH of a 1.0 mol dm^{-3} solution of hydrogen cyanide at 25°C. **(3 marks)**

(ii) State and justify any assumptions that you made. **(2 marks)**

c Calculate the ratio of [CN^-]:[HCN] in a buffer solution of hydrogen cyanide and potassium cyanide that is pH 9. **(3 marks)**

d Hydrogen cyanide reacts with ethanal at a pH between 8 and 10.

$$HCN + CH_3CHO \rightarrow CH_3CH(OH)CN$$

The initial rates of this reaction with different buffer solutions of HCN and KCN were determined. The results are shown in the table below.

Experiment	[CH_3CHO] / mol dm^{-3}	[CN^-]/mol dm^{-3}	Initial rate/ mol dm^{-3} s^{-1}
1	0.10	0.20	2.4×10^{-4}
2	0.10	0.40	4.8×10^{-4}
3	0.40	0.40	1.9×10^{-3}

(i) What is the order with respect to CH_3CHO and to CN^-? Justify your answers. **(3 marks)**

(ii) Write a mechanism for this reaction that is consistent with the orders you found in **d (i)**. **(3 marks)**

(iii) Explain why the product of this reaction has no effect on the plane of polarisation of plane-polarised light. **(2 marks)**

Total: 18 marks

23 Acrolein, CH_2–CHCHO, is one of the causes of the smell of barbecued meat and smoky bacon.

a Describe how acrolein can be converted into propenoic acid, CH_2=CHCOOH, in the laboratory. **(3 marks)**

b Explain why propenoic acid has a higher boiling point than acrolein. **(2 marks)**

c Write the structural formulae of the organic product of reacting propenoic acid with the following:

(i) lithium aluminium hydride (lithium tetrahydridoaluminate(III)) **(1 mark)**

(ii) hydrogen in the presence of a nickel catalyst **(1 mark)**

d Write the equations for the reactions of propenoic acid with:

(i) phosphorus pentachloride **(1 mark)**

(ii) sodium carbonate. **(1 mark)**

e The NMR spectrum of acrolein shows a doublet at δ = 9.1; that of propenoic acid has a singlet at δ = 12.0. Suggest which hydrogen nuclei in these compounds are responsible for these peaks and explain why one peak is split into two and the other is not split. **(4 marks)**

f Propanedioic acid, $HOOCCH_2COOH$, can form a polyester with propane-1,3-diol, $HOCH_2CH_2CH_2OH$. Draw the repeat unit of this polyester. **(2 marks)**

Total: 15 marks

Section B total: 50 marks

Section C

24 This question concerns an organic compound **X**.

a Analysis of **X** showed that it contained 66.7% carbon, 11.1% hydrogen and 22.2% oxygen. Show that this is consistent with **X** having an empirical formula of C_4H_8O. **(2 marks)**

b The mass spectrum of **X** had a molecular ion peak at m/e = 72. Deduce the molecular formula of **X**. **(2 marks)**

c State three functional groups that could be present in **X**. **(3 marks)**

d Several tests were carried out on **X**. The results are shown in **(i)** to **(v)** below. What can be concluded about **X** from each of these tests?

(i) When bromine water was added, there was no colour change. **(1 mark)**

(ii) When phosphorus pentachloride was added, no steamy fumes were observed. **(1 mark)**

(iii) When Brady's reagent was added, a red precipitate was observed. **(1 mark)**

(iv) No precipitate was seen when **X** was warmed with a solution of iodine and alkali. **(1 mark)**

(v) A silver mirror was observed when **X** was warmed with Tollens' reagent. **(1 mark)**

e Write displayed formulae of two compounds that could be **X**. **(2 marks)**

f **(i)** The mass spectrum showed a line at m/e = 29. Identify one species that could be responsible for this line. **(1 mark)**

(ii) The NMR spectrum of **X** had four peaks. Use this fact to deduce which of the two structures that you drew in **e** is compound **X**. Justify your answer. **(2 marks)**

g Given the following data, calculate the enthalpy of formation of **X**. **(3 marks)**

	Compound **X**, $C_4H_8O(l)$	C(s)	$H_2(g)$
$\Delta H_{combustion}$/kJ mol^{-1}	–2476	–394	–286

Total: 20 marks

Paper total: 90 marks

UNIT 5

Transition metals and organic nitrogen chemistry

Topic 5.1

Redox equilibria

Introduction

- ★ You must revise redox in Unit 2 (page 37).
- ★ The sign of $E^\ominus$ indicates the direction of spontaneous reaction.
- ★ A useful mnemonic is **oil rig** — when a substance is **o**xidised **i**t **l**oses one or more electrons; when **r**educed **i**t **g**ains one or more electrons.
- ★ A half-equation always includes electrons.
- ★ Half-equations are usually given as reduction potentials, i.e. with the electrons on the *left*-hand side.
- ★ An oxidising agent becomes reduced when it reacts.

Things to learn

- ★ The potential of the standard hydrogen electrode at 25°C is zero.
- ★ The standard electrode potential is the potential of a half-cell relative to that of a standard hydrogen electrode. The concentration of all ions in solution is 1.0 mol dm^{-3}, all gases are at 1 atm pressure and the system is at a stated temperature, usually 25°C (298 K).
- ★ A reaction is feasible (thermodynamically favourable) if E_{cell} is positive.

Things to understand

The standard hydrogen electrode

This consists of hydrogen gas at 1 atm pressure bubbling over a platinum electrode immersed in a 1 mol dm^{-3} solution of H^+ ions, at a temperature of 25°C (see Figure 5.1).

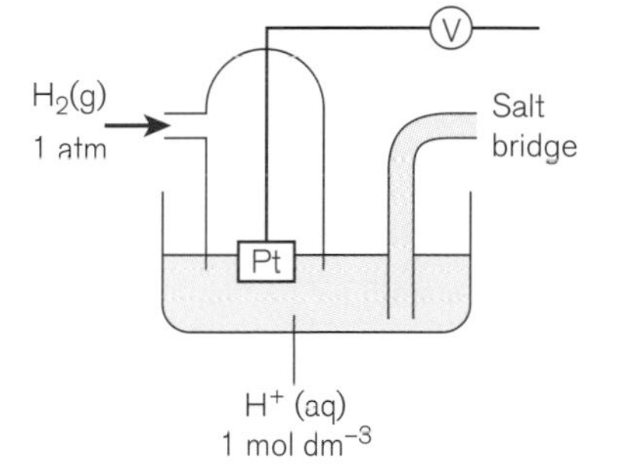

Figure 5.1 A standard hydrogen electrode

Standard electrode potential, symbol $E^\ominus$

- ★ For a **metal**, it is when the metal is immersed in a solution of its ions at a concentration of 1 mol dm^{-3}. For example, for the reaction:

  $$Zn^{2+}(aq) + 2e^- \longrightarrow Zn(s)$$

  it is for a zinc electrode immersed in a 1 mol dm^{-3} solution of Zn^{2+} ions at 25°C.
- ★ For a **non-metal**, it is when the non metal at 1 atm pressure (or a solution of it at a concentration of 1 mol dm^{-3}) is in contact with a platinum electrode immersed in a solution of the ions of the non-metal at a concentration of 1 mol dm^{-3} at 25°C. For example, for the reaction:

  $$\tfrac{1}{2}Cl_2(g) + e^- \longrightarrow Cl^-(aq)$$

  it is for chlorine gas, at 1 atm pressure, being bubbled over a platinum electrode dipping into a 1 mol dm^{-3} solution of Cl^- ions at 25°C.

- ★ For a **redox system of two ions of an element**, it is when a platinum electrode is immersed in a solution containing all the ions in the half-equation at a concentration of 1 mol dm^{-3}. For example, for the reaction:

  $Fe^{3+}(aq) + e^{-} \longrightarrow Fe^{2+}(aq)$

  it is for a platinum electrode immersed in a solution which is 1 mol dm^{-3} in both Fe^{3+} and Fe^{2+} ions at 25°C.
- ★ The equations are always given as **reduction potentials**, i.e. with the electrons on the *left*-hand side.
- ★ Comparing substances on the *left*-hand side of two half-equations, the one with the *more* positive (or less negative) value of $E^{\ominus}$ is the more powerful oxidising agent. (It is the most easily reduced).
- ★ Comparing substances on the *right*-hand side of two half-equations, the one with the *less* positive (or more negative) value of $E^{\ominus}$ is the more powerful reducing agent.

$\frac{1}{2}Cl_2 + e^{-} \rightleftharpoons Cl^{-}$ $E^{\ominus} = +1.36$ V
$\frac{1}{2}Br_2 + e^{-} \rightleftharpoons Br^{-}$ $E^{\ominus} = +1.07$ V
As +1.36 > +1.07, chlorine is a stronger oxidising agent than bromine.

$Sn^{4+} + 2e^{-} \rightleftharpoons Sn^{2+}$ $E^{\ominus} = +0.15$ V
$Fe^{3+} + e^{-} \rightleftharpoons Fe^{2+}$ $E^{\ominus} = +0.77$ V
As +0.15 < +0.77, Sn^{2+} is a better reducing agent than Fe^{2+}.

Calculation of $E^{\ominus}_{cell}$ and the overall equation

$E^{\ominus}_{cell}$ from half-equations

First, identify the reactants. One will be on the left-hand side of a reduction half-equation; the other will be on the right-hand side of a reduction half-equation. The half-equation with the reactant on the right has to be *reversed* and this process *alters the sign* of $E^{\ominus}$. The two $E^{\ominus}$ values are then added giving $E^{\ominus}_{cell}$. If this is a positive number, the reaction is feasible.

Overall equation from half-equations

The number of electrons in each half-equation must be the same, so one or both must be multiplied by an integer until this is attained.

When a redox half-equation is reversed, the sign of $E^{\ominus}$ must be changed.
When a redox half-equation is multiplied, the $E^{\ominus}$ value is not altered.

Worked example

Use the following data to deduce the overall equation and the value of $E^{\ominus}_{cell}$ for the reaction between acidified potassium manganate(VII) and iron (II) ions:

(i) $MnO_4^{-}(aq) + 8H^{+}(aq) + 5e^{-} \rightleftharpoons Mn^{2+}(aq) + 4H_2O(l)$ $E^{\ominus} = +1.52$ V
(ii) $Fe^{3+}(aq) + e^{-} \rightleftharpoons Fe^{2+}(aq)$ $E^{\ominus} = +0.77$ V

Answer

Reverse equation (ii) and multiply it by 5. Then add it to equation (i).

$5Fe^{2+}(aq) \rightleftharpoons 5Fe^{3+}(aq) + 5e^{-}$ $E^{\ominus} = -(+0.77\ V) = -0.77$ V
$MnO_4^{-}(aq) + 8H^{+}(aq) + 5e^{-} \rightleftharpoons Mn^{2+}(aq) + 4H_2O(l)$ $E^{\ominus} = +1.52$ V

$MnO_4^{-}(aq) + 8H^{+}(aq) + 5Fe^{2+}(aq) \rightleftharpoons Mn^{2+}(aq) + 4H_2O(l) + 5Fe^{3+}(aq)$

$E^{\ominus}_{cell} = -0.77 + 1.52 = +0.75$ V

Hint

Always ensure that the reactants that are specified in the question (here they are iron(II) and manganate(VII) ions) are on the left-hand side of the final overall equation.

Stoichiometry of an overall equation

The total increase in oxidation number of one element must equal the total decrease in oxidation number in another.

In the worked example above, the oxidation number of manganese decreases by 5 from +7 to +2 and that of iron increases by 1. Therefore, there must be five Fe(II) ions to each MnO_4^{-} ion in the overall equation.

As $E^\ominus_{cell}$ is proportional to ΔS_{total} and hence to ln K, the more positive $E^\ominus_{cell}$, the more complete the reaction and the greater the value of K. As a guide, a reaction involving the transfer of two electrons, with a value of $E^\ominus_{cell} > 0.1$ V, will be well to the right. For a five-electron transfer, the value required would be > 0.03 V.

Spontaneous change

- Electrode potential data can be used to predict the feasibility of a chemical reaction. A reaction is feasible (thermodynamically favourable) if E_{cell} is positive.
- However, if the **activation energy** is high, the rate of the reaction may be so slow that the reaction does not occur (kinetically stable).
- **Non-standard conditions** may result in a reaction taking place even if the standard electrode potential is negative. For example, the reaction:

$$2Cu^{2+}(aq) + 2I^-(aq) \rightleftharpoons 2CuI(s) + I_2(aq)$$

should not occur because $E^\ominus$ (assuming all species are soluble) is -0.39 V. However copper(I) iodide is precipitated and this makes $[Cu^+]$ very much less than 1 mol dm^{-3}. The equilibrium is driven to the right by the removal of Cu(I) ions, so E_{cell} becomes positive and the reaction takes place.

Potassium manganate(VII) titrations: estimation of reducing agents

- Acidified potassium manganate(VII) will quantitatively oxidise many reducing agents.
- The procedure is to pipette a known volume of the reducing agent into a conical flask and add an excess of dilute sulfuric acid.
- A potassium manganate(VII) solution of known concentration is placed in a burette and run in until a faint pink colour appears.
- This shows that there is a minute excess of the manganate(VII) ions.
- If the stoichiometry of the reaction is known, the concentration of the reducing agent can be calculated. No indicator is needed because manganate(VII) ions are intensely coloured.

Worked example

25.0 cm^3 of a solution of iron(II) sulfate was acidified and titrated against 0.0222 mol dm^{-3} potassium manganate(VII) solution. 23.4 cm^3 were required to give a faint pink colour. Calculate the concentration of the iron(II) sulfate solution.

Answer

The equation for the reaction is:

$$MnO_4^-(aq) + 8H^+(aq) + 5Fe^{2+}(aq) \longrightarrow Mn^{2+}(aq) + 4H_2O(l) + 5Fe^{3+}(aq)$$

amount of manganate(VII) = $0.0222 \times 23.4/1000 = 5.195 \times 10^{-4}$ mol

amount of iron(II) sulfate = $5.195 \times 10^{-4} \times 5/1 = 2.597 \times 10^{-3}$ mol

concentration of iron(II) sulfate = $2.597 \times 10^{-3}/0.0250 = 0.104$ mol dm^{-3}

The ratio of iron(II) to manganate(VII) ions in the equation is 5:1, so the number of moles of Fe^{2+} is five times the number of moles of manganate(VII).

Iodine–thiosulfate titrations: estimation of oxidising agents

- The procedure is to add a 25.0 cm^3 sample of an oxidising agent to excess potassium iodide solution (often in the presence of dilute sulfuric acid).
- The oxidising agent liberates iodine, which can then be titrated against standard sodium thiosulfate solution. When the iodine colour becomes pale straw, starch indicator is added and the addition of sodium thiosulfate is continued until the blue colour disappears.
- Iodine reacts with thiosulfate ions according to the equation:

$$I_2 + 2S_2O_3^{2-} \longrightarrow 2I^- + S_4O_6^{2-}$$

Worked example

$25.0\ cm^3$ of a solution of hydrogen peroxide, H_2O_2, was added to excess acidified potassium iodide solution. The liberated iodine required $23.8\ cm^3$ of $0.106\ mol\ dm^{-3}$ sodium thiosulfate solution. Calculate the concentration of the hydrogen peroxide solution.

Answer

The equation for the oxidation of iodide ions by hydrogen peroxide is:

$$H_2O_2 + 2H^+ + 2I^- \longrightarrow I_2 + 2H_2O$$

The ratio of iodine to thiosulfate ions is 1:2, so the number of moles of iodine is half the number of moles of thiosulfate.

amount of sodium thiosulfate = $0.106 \times 23.8/1000 = 2.523 \times 10^{-3}$ mol
amount of iodine produced = $2.523 \times 10^{-3} \times ½ = 1.261 \times 10^{-3}$ mol
amount of hydrogen peroxide = $1.261 \times 10^{-3} \times 1/1 = 1.261 \times 10^{-3}$ mol
concentration of H_2O_2 = $1.261 \times 10^{-3}/0.0250 = 0.0505\ mol\ dm^{-3}$

Disproportionation

A disproportionation reaction can be predicted by using electrode potentials. Use the data to work out E_{cell} for the proposed disproportionation reaction. If it is positive, the reaction will occur.

Worked example

Will copper(I) ions disproportionate in aqueous solution to copper metal and copper(II) ions?

Data: $Cu^+(aq) + e^- \rightleftharpoons Cu(s)$ $\quad E^\ominus = +0.52\ V$
$Cu^{2+}(aq) + e^- \rightleftharpoons Cu^+(aq)$ $\quad E^\ominus = +0.15\ V$

Answer

Reverse the second equation and add it to the first. This gives the equation for the disproportionation reaction.

$$2Cu^+(aq) \rightleftharpoons Cu^{2+}(aq) + Cu(s) \qquad E^\ominus_{cell} = +0.52 - (+0.15) = +0.37\ V$$

As $E^\ominus_{cell}$ is positive, the reaction is feasible. Therefore, aqueous copper(I) ions should disproportionate.

Fuel cells

A fuel cell is an electrochemical device that combines hydrogen (or an alcohol) and oxygen (from the air) to produce electricity, with water and heat as the only by-products. As long as fuel is supplied and the by-products are removed, the fuel cell will continue to generate power. A fuel cell is two to three times more efficient than a car engine.

In a modern fuel cell hydrogen flows into a platinum coated carbon nanotube anode where it is oxidised:

$$H_2 \longrightarrow 2H^+ + 2e^-$$

A hydrogen fuel cell does not have a zero-carbon footprint because hydrocarbon fuels are used in the manufacture of hydrogen.

The hydrogen ions travel through the solid polymer electrolyte to the cathode where the oxygen is reduced.

$$½O_2 + 2H^+ + 2e^- \longrightarrow H_2O$$

Breathalysers

- ★ The oldest type consisted of a tube of potassium dichromate(VI) on silica gel. The alcohol in the breath reduced the orange gel to give a green colour. The alcohol content of breath is measured by the extent of this colour change.
- ★ A second type consisted of a spectrometer that measured the intensity of infrared absorption in the fingerprint region. This is more accurate than the dichromate test, but is still not very accurate.
- ★ The modern version is a fuel cell in which the ethanol is oxidised to ethanoic acid at the anode with oxygen from air being reduced at the cathode.
- ★ The greater the amount of alcohol in the breath sample that is oxidised, the more free electrons are produced, which results in more electricity being produced.

Checklist

Before attempting the questions on this topic, check that you can:

- ☐ define standard electrode potential
- ☐ describe the construction of a standard hydrogen electrode
- ☐ write half-equations and use them to deduce overall equations
- ☐ predict the feasibility and extent of redox and disproportionation reactions
- ☐ deduce oxidation numbers and use them to balance redox equations
- ☐ recall the principles of manganate(VII) and thiosulfate titrations
- ☐ recall the principles of fuel cells

Testing your knowledge and understanding

For the following set of questions, cover the margin, write your answers, then check to see if you are correct.

★ **a** Write the equation representing the reaction that takes place in a standard hydrogen electrode.

b What are the conditions, other than temperature, for this electrode?

★ What are the oxidation numbers of manganese in MnO_4^- and in Mn^{2+}?

★ What are the oxidation numbers of sulfur in SO_3^{2-} and in SO_4^{2-}?

★ What is the ratio of SO_3^{2-} to MnO_4^- in the reaction between them?

★ Why it is not necessary to have an indicator present in potassium manganate(VII) titrations?

Answers

a $\frac{1}{2}H_2(g) \rightleftharpoons H^+(aq) + e^-$

Hint

State symbols are essential.

b $H_2(g)$ at 1 atm pressure, $[H^+] = 1\,mol\,dm^{-3}$

+7 and +2

+4 and +6

5:2; so that in the balanced equation the total oxidation number change of both is 10

Because the manganate(VII) ions are intensely coloured.

The **answers** to the **numbered questions** are on page 145

1 a Write ionic half-equations for the following reductions:

(i) $Cr_2O_7^{2-}$ to Cr^{3+} in acidic solution. $E^\ominus = +1.33\,V$

(ii) Sn^{4+} to Sn^{2+} $E^\ominus = +0.15\,V$

(iii) iodate(V) ions, IO_3^-, to I_2 in acidic solution $E^\ominus = +1.19\,V$

(iv) I_2 to I^- $E^\ominus = +0.54\,V$

b Write overall ionic equations, calculate E_{cell} values and hence comment on the feasibility of the reactions between:

(i) potassium dichromate(VI) and tin(II) chloride in acid solution

(ii) potassium iodate(V) and potassium iodide in acid solution

2 1.320 g of mild-steel filings were reacted with excess dilute sulfuric acid. The resulting solution, containing $Fe^{2+}(aq)$ ions, was made up to a volume of 250 cm^3. 25.0 cm^3 samples of this were titrated against 0.0200 mol dm^{-3} potassium manganate(VII) solution. The mean titre was 23.5 cm^3. Calculate the percentage of iron in the steel.

3 Use the data below to answer the questions that follow.

Half-equation	E°/V
$Cr^{3+}(aq) + 3e^- \rightleftharpoons Cr(s)$	−0.90
$Zn^{2+}(aq) + 2e^- \rightleftharpoons Zn(s)$	−0.76
$Cr^{3+}(aq) + e^- \rightleftharpoons Cr^{2+}(aq)$	−0.41

a For the reduction of Cr^{3+} ions to Cr^{2+} ions by zinc, write the equation and calculate the E°_{cell} value.

b For the reduction of Cr^{3+} ions to chromium metal by zinc, write the equation and calculate the E°_{cell} value.

c Use your answers to **a** and **b** to state, with a reason, what the final products would be when zinc and Cr^{3+}ions are mixed.

Topic 5.2 Transition metal chemistry

Introduction

★ You should know the colour of the aqua complex ions and of the hydroxides of the *d*-block elements scandium to zinc.

★ You must be able to link the reactions in this topic to the theory of redox equilibria, where appropriate.

Things to learn

Scandium and zinc are not transition metals even though they are in the *d*-block. This is because their ions have the electron structure [Ar] $3d^0$ $4s^0$ and [Ar] $3d^{10}$ $4s^0$ respectively.

★ A **transition element** has at least one ion with a partly filled *d*-shell.

★ **Electron structure of the atoms**:

Note that the number of *d* electrons increases left to right, except that Cr and Cu have $4s^1$ electron structures, whereas the others have $4s^2$.This is because stability is gained when the *d*-shell is either half-full or full.

	Sc	Ti	V	Cr	Mn	Fe	Co	Ni	Cu	Zn
3*d*	1	2	3	**5**	5	6	7	8	**10**	10
4*s*	2	2	2	**1**	2	2	2	2	**1**	2

★ **Electron structure of the ions:** when forming an ion the element first loses its 4*s*-electrons. Thus the electron structures of chromium and its ions are:

Cr [Ar] $3d^5$ $4s^1$
Cr^{2+} [Ar] $3d^4$, $4s^0$
Cr^{3+} [Ar] $3d^3$ $4s^0$

★ **Successive ionisaton energies:** the energy levels of the 3*d*- and the 4*s*-electrons are very similar in *d*-block elements, so the big jump comes after all the 4*s*- and 3*d*-electrons have been removed. There is no big jump between removing the 4*s*-electrons and starting to remove the 3*d*-electrons. For chromium, there is a big jump after removing six electrons because the seventh has to come from a 3*p*-orbital.

Things to understand

Properties of transition metals

Variable oxidation state

Hydrated cations of charge greater than or equal to +4 do not exist because the sum of all the ionisation energies is too large to be compensated for by the hydration energies.

★ Formation of cations: the increase between successive ionisation energies is compensated for by a similar increase in hydration energies. Thus cations in different oxidation states are energetically favourable for all transition metals. Copper forms Cu^+ and Cu^{2+} ions; chromium forms Cr^{2+} and Cr^{3+} ions.

★ **Formation of covalent bonds**: transition metals can make available a variable number of *d*-electrons for covalent bonding. The energy required for the promotion of an electron from a 3*d*-orbital to a higher energy orbital is compensated for by the bond energy released.
Transition metals form oxo-ions. In ions such as CrO_4^{2-}, MnO_4^-, VO_3^-, VO_2^+ and VO^{2+}, the oxygen is covalently bonded to the transition metal which provides a varying number of *d*–electrons.
Chromium exists in the +2 state as Cr^{2+}, in the +3 state (as in $Cr_2(SO_4)_3$ and $[Cr(H_2O)_6]^{3+}$) and in the +6 state (as in CrO_4^{2-} and $Cr_2O_7^{2-}$).
Copper exists in the +1 state (as in CuCl, $[CuCl_2]^-$ and Cu_2O) and in the +2 state (as in $CuSO_4$ and $[Cu(NH_3)_4(H_2O)_2]^{2+}$).

Complex ion formation

★ **Bonding in complex ions:** the simplest way to view this is that the ligands form dative covalent bonds by donating a lone pair of electrons into empty orbitals of the transition metal ion.

★ **Monodentate, bidentate and polydentate ligands**
A **monodentate ligand** uses one lone pair of electrons. Typical monodentate ligands include:
- neutral molecules: H_2O (aqua), NH_3 (ammine)
- anions: F^- (fluoro), Cl^- (chloro), $(CN)^-$ (cyano)

A **bidentate** ligand uses two lone pairs from different atoms in the ligand. Typical bidentate ligands include:
- neutral molecules: $NH_2CH_2CH_2NH_2$ (1,2-diaminoethane or 'en')
- anions: $^-OOCCOO^-$ (ethanedioate or 'ox')

The **polydentate** ligand EDTA, $(^-OOCCH_2)_2NCH_2CH_2N(CH_2COO^-)_2$, is a hexadentate ligand.

The coordination number is the number of dative bonds from the ligands to the metal ion.

★ **Shapes of complex ions**: if the complex ion has six lone pairs donated by ligands such as for $[Cr(NH_3)_6]^{3+}$ and $[Cu(H_2O)_6]^{2+}$, the complex is **octahedral.** This is so that the electron pairs in the bonds to the ligands are as far apart from each other as possible.
$[CrCl_4]^-$ is tetrahedral, $[PtCl_2(NH_3)_2]$ is square planar and $[CuCl_2]$ is linear.

Coloured complex ions

★ If the ion has partially filled *d*-shell, it will be coloured. Sc^{3+} and Ti^{4+} have d^0 structures, and Cu^+ and Zn^{2+} have d^{10} structures and so are not coloured.

★ In octahedral complexes, the ligands cause splitting of the *d*-orbitals into two of higher energy and three of lower energy.

★ When white light passes through the substance, a *d*-electron is moved from a lower energy to a higher energy level.

- The frequency of the light that causes this jump is in the visible range, so that colour of this frequency is removed from the white light.
- The colour depends on the size of the energy gap, which varies with the metal ion and with the type of ligand.

> $[Cu(H_2O)_6]^{2+}$ is turquoise blue, $[Cu(NH_3)_4(H_2O)_2]^{2+}$ is deep blue, and $[CuCl_4]^{2-}$ is yellow.

Catalytic activity

Transition metals and their compounds are often good catalysts.

- Vanadium(v) oxide is used in the oxidation of SO_2 to SO_3 in the Contact process for the manufacture of sulfuric acid.
- Iron is used in the Haber process for the manufacture of ammonia.
- Nickel is used in the addition of hydrogen to alkenes.

Reactions of transition metal compounds

Reactions with sodium hydroxide and ammonia solutions

Aqueous ion	Colour	Addition of NaOH(aq)	Excess NaOH(aq)	Addition of NH_3(aq)	Excess NH_3(aq)
Cr^{3+}	Green	Green precipitate	Green solution	Green precipitate	Precipitate remains
Mn^{2+}	Pale pink	Buff precipitate[a]	Precipitate remains	Sandy precipitate[a]	Precipitate remains
Fe^{2+}	Pale green	Dirty green precipitate[b]	Precipitate remains	Dirty green precipitate[b]	Precipitate remains
Fe^{3+}	Brown/yellow	Foxy red precipitate	Precipitate remains	Foxy red precipitate	Precipitate remains
Ni^{2+}	Green	Green precipitate	Precipitate remains	Green precipitate	Green solution
Cu^{2+}	Pale blue	Blue precipitate[c]	Precipitate remains	Pale blue precipitate	Deep blue solution
Zn^{2+}	None	White precipitate	Colourless solution	White precipitate	Colourless solution

a The precipitate of $Mn(OH)_2$ goes brown as it is oxidised by air to MnO_2.
b The precipitate of $Fe(OH)_2$ goes brown on the surface as it is oxidised by the oxygen in the air to $Fe(OH)_3$.
c The precipitate of $Cu(OH)_2$ goes black as it loses water to form CuO.

Deprotonation

- The aqua ions in solution are partially deprotonated by water. The greater the surface charge density of the ion the greater the extent of this reaction, e.g. hexa-aqua chromium(III) ions:

  $$[Cr(H_2O)_6]^{3+}(aq) + H_2O(l) \rightleftharpoons [Cr(H_2O)_5(OH)]^{2+}(aq) + H_3O^+(aq)$$

  This means that solutions of chromium(III) ions are acidic (pH < 7).
- When an alkali such as sodium hydroxide is added, the equilibrium is driven to the right, the ion is considerably deprotonated to form a neutral molecule which loses water to form a precipitate of the metal hydroxide:

  $$[Cr(H_2O)_6]^{3+}(aq) + 3OH^-(aq) \rightarrow Cr(OH)_3(s) + 6H_2O$$

  If aqueous ammonia is added, the same product is formed:

  $$[Cr(H_2O)_6]^{3+}(aq) + 3NH_3(aq) \rightarrow Cr(OH)_3(s) + 3NH_4^+(aq) + 3H_2O$$

> Amphoteric hydroxides react with excess strong alkali *and* with acids,
> e.g. $Cr(OH)_3(s) + 3OH^-(aq) \rightarrow [Cr(OH)_6]^{3-}(aq)$
> $Cr(OH)_3(s) + 3H^+(aq) \rightarrow Cr^{3+}(aq) + 3H_2O(l)$

Ligand exchange

When aqueous ammonia is added to aqua complexes of *d*-block elements such as those of nickel, copper and zinc, ligand exchange takes place and a solution of the ammine complex is formed:

★ First the hydroxide is precipitated in a **deprotonation** reaction:

$[Cu(H_2O)_6]^{2+}(aq) + 2NH_3(aq) \rightarrow Cu(OH)_2(s) + 2NH_4^+(aq) + 4H_2O$

★ The hydroxide then ligand exchanges to form an ammine complex with excess ammonia:

$Cu(OH)_2(s) + 4NH_3(aq) + 2H_2O(l) \rightarrow [Cu(NH_3)_4(H_2O)_2]^{2+} + 2OH^-(aq)$

The final result is that the NH_3 ligand takes the place of the H_2O ligand.

This reaction is used as a test for:

a nickel(II): with aqueous ammonia, nickel(II) ions first give a green precipitate which then forms a blue solution with excess ammonia

b copper(II): with excess ammonia, copper(II) ions give a pale blue precipitate which then forms a deep blue solution

c zinc ions: with excess ammonia, zinc ions give a white precipitate which then forms a colourless solution

★ **Entropy of ligand exchange**

Ligand exchange takes place for three reasons:

(1) The new ligand-to-metal ion bond is stronger than the original bond.

(2) There is a very high concentration of the reacting ligand:

$[Cu(H_2O)_6]^{2+} + 4Cl^- \rightleftharpoons [CuCl_4]^{2-} + 6H_2O$

Concentrated HCl will drive this equilibrium to the right.

(3) There is an increase in entropy due to ligand exchange. This will happen when a bidentate ligand, such as diaminoethane, en, or a polydentate ligand such as EDTA, replaces monodentate ligands:

$[Cu(H_2O)_6]^{2+}(aq) + 3en(aq) \rightarrow [Cu(en)_3]^{2+}(aq) + 6H_2O(l)$

Four particles become seven particles, so entropy increases.

$[Cu(H_2O)_6]^{2+}(aq) + EDTA^{4-}(aq) \rightarrow [CuEDTA]^{2-}(aq) + 6H_2O(l)$

Two particles become seven particles, so entropy increases.

Disproportionation

See page 39.

Redox reactions of chromium

Hint

Remember that the reactants must be on the left-hand side of the overall equation. To achieve this, one half-reduction equation must be reversed and the sign of $E^\ominus$ changed. In this example, it is the sign of $E^\ominus$ for Cr^{3+}/Cr^{2+} that must be changed.

★ **Chromium(II) to chromium(III)**: any oxidising agent whose reduction potential is less negative than −0.41 V (or is positive) will oxidise Cr^{2+} to Cr^{3+}, e.g. oxygen in the air (O_2/H_2O, $E^\ominus = +1.23\,V$).

$Cr^{3+} + e^- \rightleftharpoons Cr^{2+}$ $\quad E^\ominus = -0.41\,V$

$\frac{1}{2}O_2 + 2H^+ + 2e^- \rightleftharpoons H_2O$ $\quad E^\ominus = +1.23\,V$

$E^\ominus_{cell} = -(-0.41) + 1.23 = +1.63\,V$

The reactants are Cr^{3+} and Zn, so the sign of $E^\ominus$ for the second half-equation must be changed.

★ **Chromium(III) to chromium(II):** any reducing agent for which the reduction potential is more negative than −0.41 V will reduce Cr^{3+} to Cr^{2+}, e.g. zinc metal (Zn^{2+}/Zn, $E^\circ = -0.76\,V$).

$Cr^{3+} + e^- \rightleftharpoons Cr^{2+}$ $\quad E^\ominus = -0.41\,V$

$Zn^{2+} + 2e^- \rightleftharpoons Zn$ $\quad E^\ominus = -0.76\,V$

$E^\ominus_{cell} = -(-0.76) + (-0.41) = +0.35\,V$

★ **Chromium(III) to chromium(VI):** any oxidising agent for which the reduction potential in alkaline solution is less negative than −0.13 V (or is positive) will oxidise Cr^{3+} to CrO_4^{2-}, e.g. hydrogen peroxide (H_2O_2/OH^-, $E^\ominus = +0.94\,V$)

$CrO_4^{2-} + 4H_2O + 3e^- \rightleftharpoons Cr(OH)_3 + 5OH^-$ $\quad E^\ominus = -0.13\,V$

$H_2O_2 + 2e^- \rightleftharpoons 2OH^-$ $\quad E^\ominus = +0.94\,V$

$E^\ominus_{cell} = -(-0.13) + 0.94 = +1.07\,V$

Checklist

Before attempting the questions on this topic, check that you:

- [ ] can define transition elements
- [ ] can write the electron structure of the *d*-block elements and their ions
- [ ] can recall the characteristic properties of the transition elements
- [ ] understand the nature of the bonding in complex ions, the shape of the ions and why they are coloured.
- [ ] can recall the action of aqueous sodium hydroxide and ammonia on solutions of the aqua ions
- [ ] understand ligand exchange reactions and their entropy changes
- [ ] know reactions to interconvert the oxidation states of chromium and calculate their $E^{\ominus}_{cell}$ values
- [ ] can recall examples of transition metals or their compounds as catalysts

Testing your knowledge and understanding

For the following set of questions, cover the margin, write your answers, then check to see if you are correct.

★ Write down the electronic structure of:
a a vanadium atom
b a V^{3+} ion

★ State the formula and colour of the aqua complex of copper(II).

★ State the formula and colour of the ammine complex of copper(II).

★ a Name the types of bonding in the aqua complex of iron(II).
b What is the shape of the iron(II) aqua complex ion?

★ State the type of reaction when aqueous sodium hydroxide is added to a solution of the aqua complex of copper(II).

★ State the type of reaction when excess aqueous ammonia is added to a solution of the aqua complex of copper(II).

★ Give an example of the use of iron as an industrial catalyst.

★ Give an industrial use of a vanadium compound as a catalyst.

1 Explain why the $[Cu(H_2O)_6]^{2+}$ ion is coloured.

2 Give the equations for the reactions resulting from the addition of small amounts of sodium hydroxide solution, followed by excess, to the following:
a a solution of the aqua complex of chromium(III)
b a solution of the aqua complex of iron(II)
c a solution of the aqua complex of zinc(II)

3 Give the equations for the reactions caused by the addition of small amounts of ammonia solution, followed by excess, to the following:
a a solution of the aqua complex of iron(III)
b a solution of the aqua complex of copper(II)

Answers

a V is [Ar] $3d^3\ 4s^2$
b V^{3+} is [Ar] $3d^2$.

$[Cu(H_2O)_6]^{2+}$; turquoise-blue.

$[Cu(NH_3)_4(H_2O)_2]^{2+}$; dark blue

a Dative covalent (ligand to ion) and covalent (within the water molecule)
b Octahedral

Deprotonation

Ligand exchange

Iron is used in the Haber process for the manufacture of ammonia.

Vanadium(V) oxide is used in the Contact process for the manufacture of sulfuric acid.

The **answers** to the **numbered questions** are on page 145

4 Explain why:

a solid hydrated copper(II) sulfate is coloured whereas both anhydrous copper(II) sulfate and the copper(I) ammonia complex are white.

b diaminomethane, $NH_2CH_2NH_2$, cannot act as a bidentate ligand whereas 1,2-diaminoethane, $NH_2CH_2CH_2NH_2$, can.

Topic 5.3 Organic chemistry: arenes

Introduction

Arenes are aromatic compounds. They are unsaturated compounds that contain a benzene ring of six carbon atoms.

Things to learn and understand

The structure and reactions of benzene

- ★ The benzene ring does *not* contain alternate double and single bonds.
- ★ All the bond lengths are the same and the molecule is planar.
- ★ There is an overlap of *p*-orbitals above and below the plane of the molecule, forming a continuous or **delocalised π-system**.
- ★ This gives stability to the benzene structure, which is why it undergoes substitution rather than addition reactions.
- ★ Benzene can be represented either by:

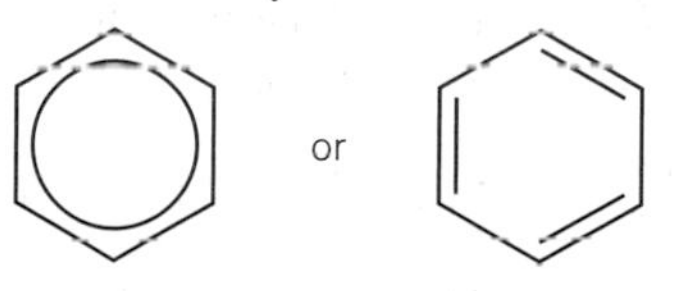

Electrophilic substitution reactions of benzene

- ★ **Nitration:** between 55°C and 60°C, benzene reacts with a mixture of concentrated nitric and sulfuric acids to form nitrobenzene:

  $C_6H_6 + HNO_3 \rightarrow C_6H_5NO_2 + H_2O$

  The mechanism of this reaction is as follows:

  Sulfuric acid is a stronger acid than nitric and so protonates it:

  $H_2SO_4 + HNO_3 \rightarrow H_2NO_3^+ + HSO_4^-$

  The cation then loses water:

  $H_2NO_3^+ \rightarrow H_2O + NO_2^+$

  The NO_2^+ ion is an electrophile and attacks the benzene ring. The HSO_4^- ion removes a H^+ reforming H_2SO_4 (the catalyst).

> Temperature control is essential. If the temperature is below 50°C the reaction is too slow; if it is above 65°C dinitrobenzene, $C_6H_4(NO_2)_2$, is formed.

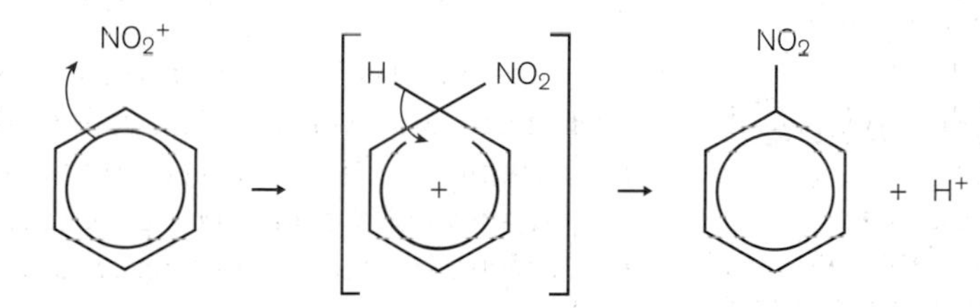

- ★ **Bromination**: benzene reacts with liquid bromine in the presence of a catalyst of anhydrous iron(III) bromide (made in situ from iron and liquid bromine) or of anhydrous aluminium bromide:

  $C_6H_6 + Br_2 \rightarrow C_6H_5Br + HBr$

The electrophile, Br^+, is formed by the reaction of bromine with the catalyst:

$Br_2 + FeBr_3 \rightarrow Br^+ + FeBr_4^-$

This then attacks the benzene ring. In the final step of the mechanism, H^+ is removed by the $FeBr_4^-$ and the catalyst is reformed.

> The reason why benzene reacts with electrophiles by substitution rather than by electrophilic addition is because the stability of the benzene ring is regained when the intermediate loses H^+.

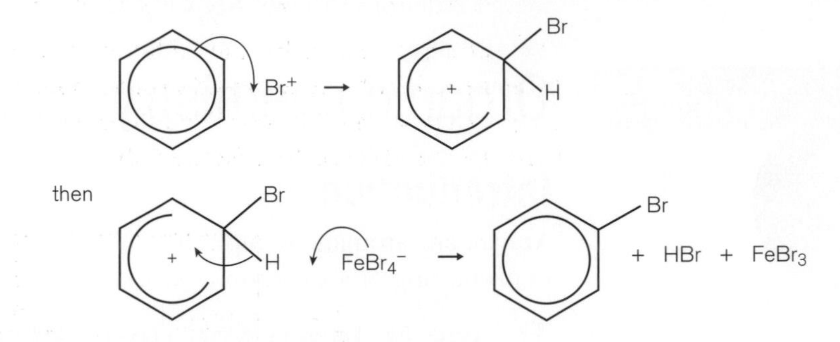

★ **Friedel–Crafts reactions:** benzene reacts with a halogenoalkane or with an acid chloride in the presence of a catalyst of anhydrous aluminium chloride. For example:

$C_6H_6 + C_2H_5Cl \rightarrow C_6H_5C_2H_5 + HCl$
ethlybenzene

$C_6H_6 + CH_3COCl \rightarrow C_6H_5COCH_3 + HCl$
phenylethanone (a ketone)

The mechanism is as follows:

$C_2H_5Cl + AlCl_3 \rightarrow CH_3CH_2^+ + AlCl_4^-$
then

> **Hint**
>
> Make sure that the curly arrow goes to the CH_2 carbon, which has a + charge, and not to the CH_3 carbon, and that the bond to the benzene ring is *not* through the CH_3 group.

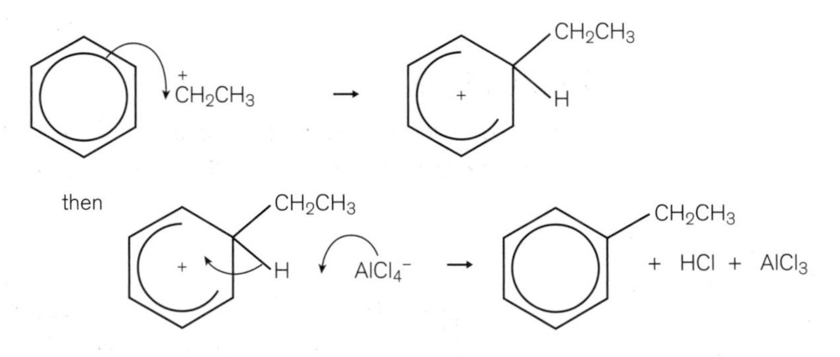

For an acidic chloride reacting with benzene, the reaction is:

$CH_3COCl + AlCl_3 \rightarrow CH_3C^+O + AlCl_4^-$ then

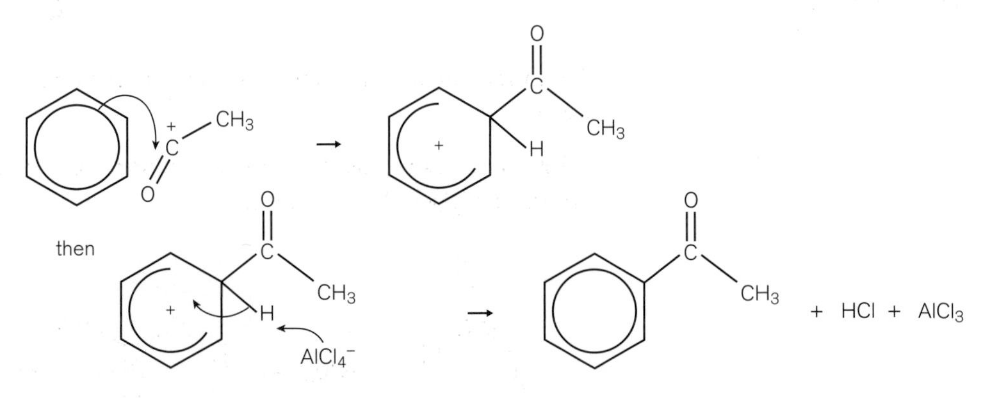

★ **Sulfonation:** when benzene is warmed with fuming sulfuric acid (SO_3 dissolved in concentrated H_2SO_4), benzenesulfonic acid is produced.

$C_6H_6 + SO_3 \rightarrow C_6H_5SO_3H$

Addition of hydrogen to benzene

When benzene and hydrogen are passed over a hot nickel catalyst, cyclohexane is formed:

$$C_6H_6 + 3H_2 \longrightarrow C_6H_{12} \qquad \Delta H = -207\,\text{kJ mol}^{-1}$$

The enthalpy change for the addition of hydrogen to cyclohexene is $-119\,\text{kJ mol}^{-1}$. If benzene had three localised double bonds, the enthalpy of hydrogenation would be $3 \times -119 = -357\,\text{kJ mol}^{-1}$. The difference of $150\,\text{kJ mol}^{-1}$ is the amount by which benzene is stabilised (see Figure 5.2).

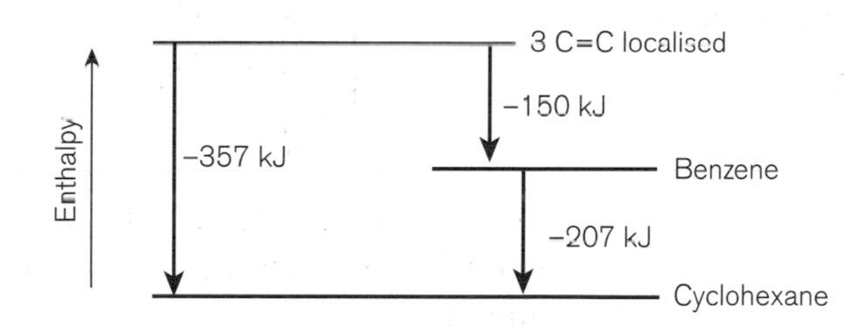

Figure 5.2 Energy-profile diagram for the addition of hydrogen to benzene

Phenol

The strucure of phenol is as follows:

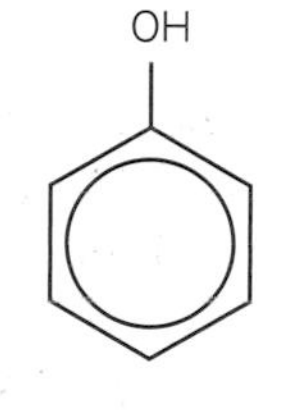

Reactions

★ With **bromine water**: the presence of the –OH group makes substitution into the benzene ring easier than with benzene. No catalyst is required.

The lone pair of p_z-electrons on the oxygen atom interacts with the delocalised π-electrons in the ring and so the electron density in the ring is increased. This means that phenol is activated towards electrophilic substitution.

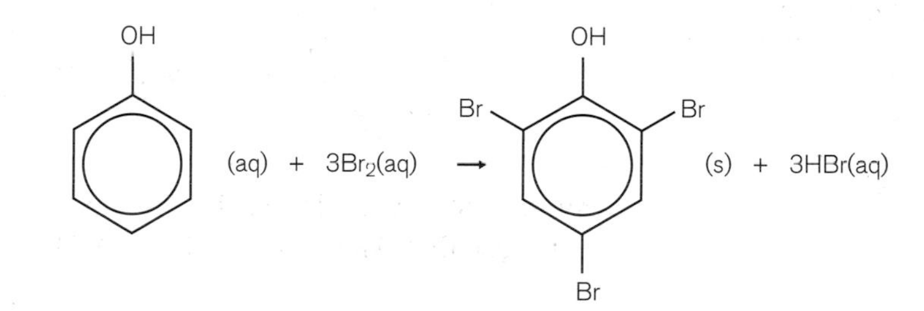

When red-brown bromine water is added to phenol, a white precipitate is rapidly formed.

★ **Nitration**: phenol reacts with dilute nitric acid to give a mixture of 2-nitrophenol and 4-nitrophenol, and water:

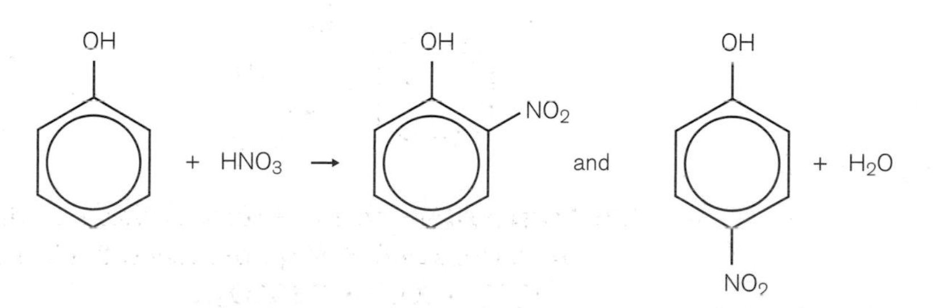

Checklist

Before attempting the questions on this topic, check that you:

- ☐ understand the structure of benzene and why its reactions are mostly substitution rather than addition
- ☐ can recall the reactions of benzene with nitric acid, bromine, sulfur trioxide and halogen compounds
- ☐ understand the mechanisms of the electrophilic substitution reactions of benzene
- ☐ can recall the reactions of phenol with bromine and dilute nitric acid

Testing your knowledge and understanding

For the following set of questions, cover the margin, write your answers, then check to see if you are correct.

- What would you observe when bromine water is added to aqueous phenol?
- Give the electrophile in:
  - **a** the nitration of benzene
  - **b** the reaction of benzene with chloroethane

1 a Describe the mechanism for the reaction of bromine with benzene.

b Why does benzene undergo a substitution reaction with bromine rather than an addition reaction?

2 Why does phenol react more easily than benzene with electrophiles?

Answers

The red-brown bromine water would become colourless and a white precipitate would be formed.

a NO_2^+

b $CH_3CH_2^+$

The **answers** to the **numbered questions** are on page 146

Topic 5.4 Organic chemistry: nitrogen compounds

Introduction

These consist of amines, nitriles, amides, amino acids and proteins.

Things to learn and understand

Primary amines

- Primary amines contain the $-NH_2$ group.
- They are soluble in water (if the carbon chain is fairly short) because they can use the lone pair of electrons on the nitrogen atom to form a hydrogen bond with a water molecule.

Secondary amines have two carbon atoms attached to an >NH group.

Reactions of ethylamine

- With **water** — amines are weak bases and produce OH^- ions when added to water:

  $$C_2H_5NH_2(aq) + H_2O(l) \rightleftharpoons C_2H_5NH_3^+(aq) + OH^-(aq)$$

- With an **acid** to form a salt:

  $$C_2H_5NH_2(aq) + H^+(aq) \longrightarrow C_2H_5NH_3^+(aq)$$

The smell of the amine disappears when acid is added and reappears when alkali is added as the amine is regenerated from the salt by the alkali.

- With an **acid chloride** to form a substituted amide:
  $C_2H_5NH_2 + CH_3COCl \rightarrow CH_3CONHC_2H_5 + HCl$
- With a **halogenoalkane** to form a secondary amine:
  $C_2H_5NH_2 + C_2H_5Cl \rightarrow (C_2H_5)_2NH + HCl$
- With **copper(II) ions** to form a complex ion (compare the complex with the ammine):
  $[Cu(H_2O)_6]^{2+} + 4C_2H_5NH_2 \rightarrow [Cu(C_2H_5NH_2)_4(H_2O)_2]^{2+} + 4H_2O$
  The colour is similar to that of the ammine complex.

Diamines, such as 1,2-diaminoethane, $H_2NCH_2CH_2NH_2$, also form a complex with copper(II) ions (see page 107).

Phenylamine

Phenylamine has an $-NH_2$ group attached to the benzene ring.

- **Preparation**: nitrobenzene is reduced by tin and concentrated hydrochloric acid when heated under reflux. Sodium hydroxide is added to liberate the phenylamine, which is removed by steam distillation.
  $C_6H_5NO_2 + 6[H] \rightarrow C_6H_5NH_2 + 2H_2O$
- Reactions of the $\mathbf{-NH_2}$ group are the same as those of ethylamine.
- Reaction with **nitrous acid** to form the diazonium ion:
  $C_6H_5NH_2 + 2H^+ + NO_2^- \rightarrow C_6H_5N_2^+ + 2H_2O$
  Dilute hydrochloric acid is added to phenylamine and the solution is cooled to 5°C. Sodium nitrite solution is then added, keeping the temperature between 0°C and 5°C.
  When the resulting solution is mixed with an alkaline solution of phenol a brightly coloured precipitate of the diazo compound is obtained:

Below 0°C the reaction is too slow; above 10°C the diazonium ion decomposes.

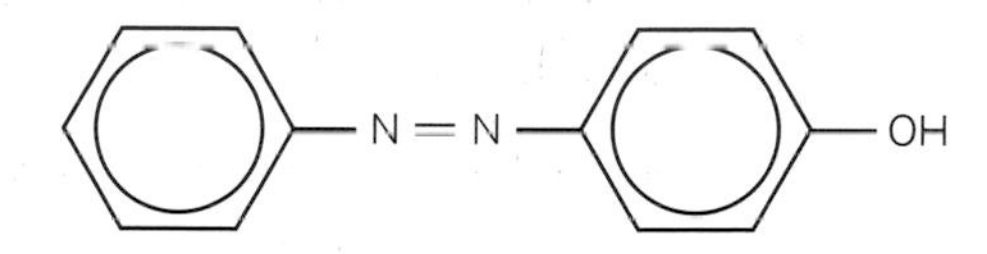

This is a coupling reaction of the diazonium ion.

Nitriles

Nitriles contain the C≡N group, e.g. propanenitrile C_2H_5CN.

Hydrolysis

- with acid:
  $C_2H_5CN + H^+ + 2H_2O \rightarrow C_2H_5COOH + NH_4^+$
  Conditions: heat under reflux with dilute sulfuric acid
- with alkali:
  $C_2H_5CN + OH^- + H_2O \rightarrow C_2H_5COO^- + NH_3$
  Conditions: heat under reflux with dilute sodium hydroxide.

Hydroxynitriles can be prepared by the addition of HCN to carbonyl compounds in the presence of a base. The –CN group reacts in the same way as nitriles.

Amides and polyamides

Amides and polyamides are prepared by the action of an acid chloride on a primary amine. To form a polyamide, the acid chloride must have two –COCl groups and the amine two $-NH_2$ groups.

- **Amides** contain the $-CONH_2$ group, e.g. ethanamide CH_3CONH_2: $CH_3-C(=O)-NH_2$
- **Polyamides** contain the $-C(=O)-N(H)-$ group which is called the peptide link.

> Nylon and other polyamides are very strong because of hydrogen bonding between an –NH group in one chain and a >C=O group in another chain.

★ **Nylon** is a polyamide that can be made from 1,2-diaminohexane, $NH_2(CH_2)_6NH_2$, and hexan-1,6-dioic acid, $HOOC(CH_2)_4COOH$ (or its diacid chloride). The repeat unit is:

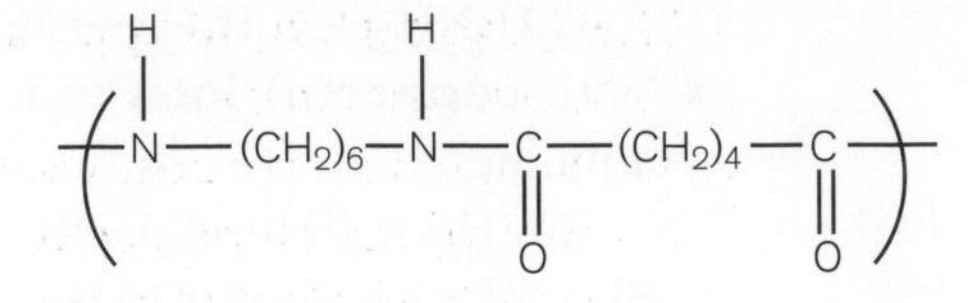

> These and the polyesters below are examples of condensation polymers. In their formation a small molecule, HCl or H_2O, is lost from the reaction between the monomers.
> Note that the repeat unit of a polyamide contains two oxygen and two nitrogen atoms.

★ **Kevlar→** is a polyamide made from 1,4-diaminobenzene and benzene-1, 4-dicarboxylic acid:

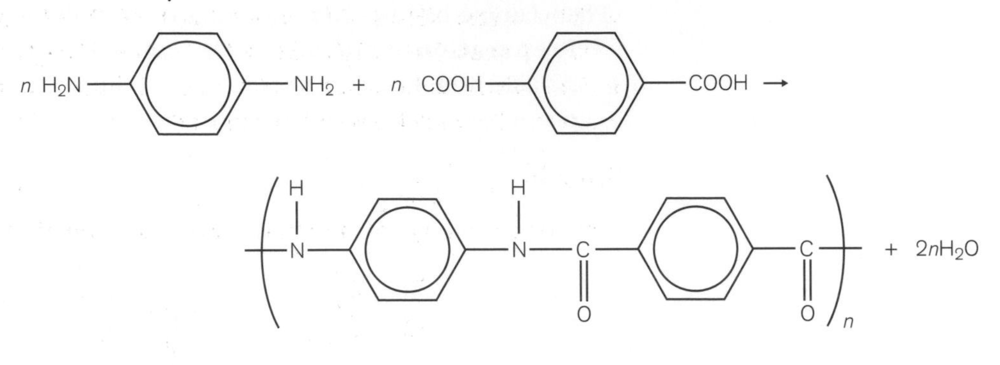

It is used in bulletproof vests and stab jackets.

Other polymers

★ **Polyesters** contain the –C(=O)–O– group. An example is Terylene®, which is made from benzene-1,4-dicarboxylic acid (or its acid chloride) and ethane-1, 2-diol. Its repeat unit is:

> Note that the repeat unit of a polyester contains four oxygen atoms.

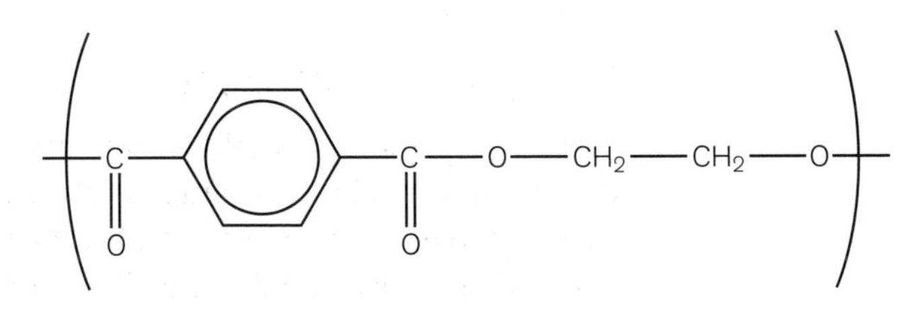

★ Addition polymers are made from compounds with a >C=C< group. Examples are poly(ethenol) and poly(propenamide).

The repeat unit of poly(ethenol) is:

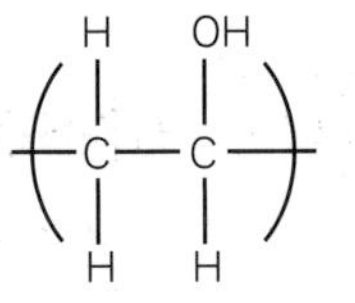

> Polyethenol is water-soluble because it has an –OH group on every other carbon atom and so can form hydrogen bonds on every –OH group in the chain.

The repeat unit of poly(propenamide) is:

Amino acids

Amino acids contain an $-NH_2$ group and a –COOH group (bonded to the same carbon atom).

- ★ They are water-soluble ionic solids because they form **zwitterions** that can form strong bonds with water molecules:
  $NH_2CH_2COOH \rightleftharpoons {}^+NH_3CH_2COO^-$
- ★ They have high melting points because strong ion–ion attractions occur between *different* zwitterions.
- ★ All natural amino acids, except aminoethanoic acid (glycine) are optically active because of the chiral centre (marked *) in the general formula $NH_2C^*HRCOOH$, where R is an alkyl or other organic group.

Reactions of aminoethanoic acid

- ★ With **acids**: $NH_2CH_2COOH + H^+(aq) \rightarrow {}^+NH_3CH_2COOH(aq)$
- ★ With **bases**: $NH_2CH_2COOH + OH^-(aq) \rightarrow NH_2CH_2COO^-(aq) + H_2O$
- ★ With **ninhydrin**: when a solution of ninhydrin is sprayed onto an amino acid and warmed, a violet colour is produced.

Proteins

Proteins are polymers of amino acids. They contain the peptide link:

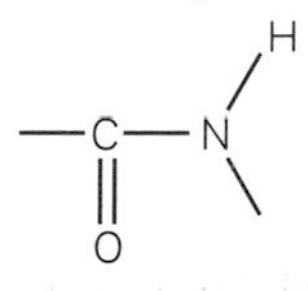

- ★ Two amino acids can form two different dipeptides:
  $NH_2CHRCOOH + NH_2CHR'COOH \rightarrow NH_2CHRCONHCHR'COOH + H_2O$
  or
  $NH_2CHRCOOH + NH_2CHR'COOH \rightarrow NH_2CHR'CONHCHRCOOH + H_2O$
- ★ Proteins can be hydrolysed to amino acids by warming with aqueous sodium hydroxide or dilute acid. These can then be identified using thin-layer chromatography, TLC.

Thin-layer chromatography

A spot of a solution containing the mixture of amino acids is placed about 2 cm from the bottom of a chromatographic plate and spots of solutions of known amino acids placed alongside. The bottom of the plate is placed in a suitable liquid eluent in a tank and left for a suitable time. The plate is then dried, warmed and sprayed with ninhydrin. If the height of one of the spots matches the height of one of the known amino acids, that amino acid must have been in the mixture.

Checklist

Before attempting questions on this topic, check that you:

- ☐ know why amines are water soluble
- ☐ can recall the reactions of amines
- ☐ know the preparation of phenylamine and its reaction with nitrous acid
- ☐ can recall the coupling reaction of diazonium ions
- ☐ can recall the hydrolysis of nitriles
- ☐ can recall the repeat units of polyamides, polyesters and poly(ethenol)
- ☐ know the physical properties and reactions of amino acids
- ☐ understand the peptide link and the method of separating mixtures of amino acids by TLC

Testing your knowledge and understanding

For the following questions, cover the margin, write your answers, then check to see if you are correct.

★ State the structural formulae of the product obtained by reacting:

- **a** phenylamine with dilute hydrochloric acid
- **b** phenylamine with ethanoyl chloride

★ State the conditions for the conversion of:

- **a** ethanenitrile, CH_3CN, to ethanoic acid, CH_3COOH
- **b** ethanenitrile, CH_3CN, to ethylamine, $C_2H_5NH_2$

★ Explain why aminoethanoic acid is a solid that is soluble in water.

Answers

- **a** $C_6H_5NH_2^+Cl^-$
- **b** $CH_3CONHC_6H_5$

- **a** Heat under reflux with NaOH(aq), then acidify
- **b** $LiAlH_4$ in dry ether, followed by dilute acid

Solid: strong ion–ion attractions between *different* zwitterions

Soluble: strong forces between charges on zwitterions and water molecules

The **answers** to the **numbered questions** are on page 146

1 Write the formulae of the products of the reactions of phenylamine with:

- **a** hydrated copper(II) ions
- **b** chloroethane
- **c** ethanoyl chloride

2 Write the structural formulae of the products of the reaction of 2-aminopropanoic acid, $CH_3CH(NH_2)COOH$, with:

- **a** H^+ ions
- **b** OH^- ions
- **c** C_2H_5OH in the presence of a concentrated sulfuric acid catalyst.

3 When benzenediazonium chloride is prepared, the reaction is carried out at about 5°C.

- **a** Why are these conditions chosen?
- **b** What is the formula of the product obtained by the reaction of the benzene diazonium chloride solution with phenol in alkaline solution?

4 Write the repeat unit of the following polymers:

- **a** Nomex, which is made from benzene-1,3-dicarboxylic acid and 1,4-diaminobenzene
- **b** poly(propenamide), which is made from propenamide, $CH_2{=}CHCONH_2$

Topic **5.5**

Organic analysis and synthesis

Introduction

This topic brings together all the organic chemistry that has been covered in the AS and A2 units.

Things to learn and understand

Organic analysis

- ★ First, find the empirical formula.
- ★ The molecular ion peak will then give the molecular formula.
- ★ Chemical tests determine the functional groups that are present.

Tests

★ **Alkenes**

Test: add bromine water

Observation: the red-brown colour of bromine is lost; the solution becomes colourless.

★ **Halogenoalkanes**

Test: heat under reflux with sodium hydroxide solution, then acidify with excess dilute nitric acid and add silver nitrate solution.

Observation: chlorides give a white precipitate that is soluble in dilute ammonia Bromides give a cream precipitate that is insoluble in dilute ammonia but dissolves in concentrated ammonia.

Iodides give a yellow precipitate that is insoluble in concentrated ammonia.

★ **–OH group (in alcohols and acids)**

Test: add phosphorus pentachloride to the dry compound.

Observation: steamy fumes are produced.

★ **Alcohols**

Test: warm with dilute sulfuric acid and potassium dichromate(VI) solution.

Observation: primary and secondary alcohols reduce the orange dichromate(VI) ions to green Cr^{3+} ions.

Tertiary alcohols do not affect the colour because they are not oxidised.

To distinguish between primary and secondary alcohols repeat the experiment, but distil the product into ammoniacal silver nitrate solution:

- primary alcohols are oxidised to aldehydes, which give a silver mirror.
- secondary alcohols are oxidised to ketones, which do not react.

★ **Carboxylic acids, COOH group**

Test: add to a solution of sodium hydrogencarbonate.

Observation: bubbles of gas, which turn limewater milky

★ **Carbonyl, C=O group** (aldehyde or ketone)

Test. add a solution of 2,4-dinitrophenylhydrazine.

Observation: a yellow or orange precipitate

★ **To distinguish between aldehydes and ketones**

Fehling's (Benedict's) solution can be used in place of Tollens' reagent. Aldehydes give a red precipitate; with ketones the solution stays blue.

Test: warm with Tollens' reagent (ammoniacal silver nitrate solution)

Observation: aldehydes give a silver mirror; ketones do not react.

★ **Iodoform** reaction

Test: warm gently with a solution of sodium hydroxide and iodine.

Observation: a pale yellow precipitate (CHI_3)

This test works with carbonyl compounds containing $CH_3C{=}O$ and alcohols containing $CH_3CH(OH)$.

Interpretation of data

You should be able to deduce structural formulae of organic molecules given data obtained from chemical methods and from spectrscopy:

★ **Chemical methods**

The solid must be purified by recrystallisation in order for it to have a sharp melting temperature to enable accurate identification.

Carry out tests as above to determine the functional groups in the molecule, then make a solid derivative (such as the 2,4-DNP derivative from carbonyl compounds), purify it, and measure its melting temperature. Check the melting temperature against values in a data bank to identify the substance.

Hint

When identifying species from lines in the spectrum, always give both the structural formula of the ion andindicate the positive charge.

★ **Mass spectra** (see Figure 5.3)
- Observe the fragments obtained and look for the value of the molecular ion, M^+, if any.
- See if there is a peak at $(M-15)^+$. If so, the substance has a CH_3 group which is lost on forming the (M–15) peak.
- If there is a peak at $(M-29)^+$, the substance has either a $-CHO$ or a $-C_2H_5$ group. Other fragments will help to identify the structure.

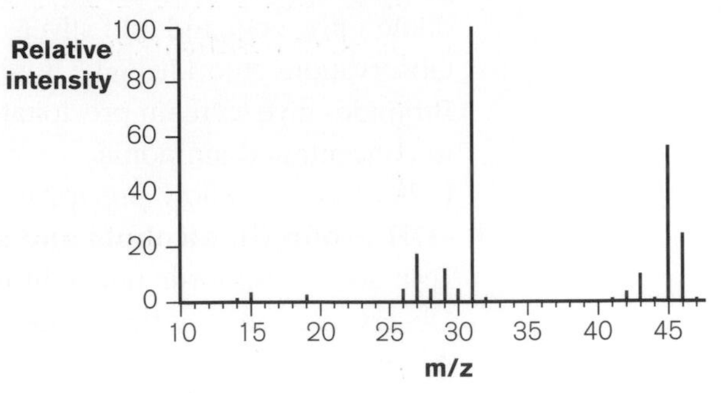

Figure 5.3 Mass spectrum of ethanol

★ **Infrared spectra** (see Figure 5.4)

Infrared spectra are used to identify functional groups (see page 61 for a fuller list). Some peaks to look for are shown in the table below.

The peaks due to O–H and N–H are usually broadened due to hydrogen bonding.

Peak range	Functional group
1650–1750 cm^{-1}	C=O (in aldehydes, ketones, acids and esters)
3200–3700 cm^{-1}	O–H (in alcohols)
3300–3500 cm^{-1}	N–H
1000–1300 cm^{-1}	C–O
2500–3300 cm^{-1}	O–H (in acids)

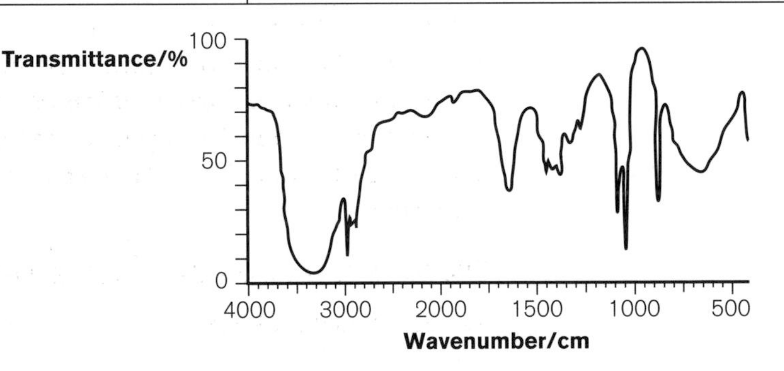

Figure 5.4 Infrared spectrum of ethanol

The chemical shift is the difference between the absorption frequencies of the hydrogen nuclei in the compound and those in the reference compound.

★ **NMR spectra**
- The nuclei of hydrogen atoms in different chemical environments within a molecule show up separately in a NMR spectrum. The values of their chemical shifts, δ, are different (see page 92).
- A hydrogen nucleus in a CH_3 group has a different chemical shift from that in a CH_2 group or an OH group.
- In **low resolution** NMR, each group shows as a single peak, and the area under the peak is proportional to the number of hydrogen atoms with the same environment. Thus ethanol, CH_3CH_2OH, has three peaks of relative intensity 3:2:1 and methyl propane, $CH_3CH(CH_3)CH_3$, has two peaks of relative intensity 9:1.

- In **high resolution** NMR, spin coupling is observed. This is caused by the interference of the magnetic fields of neighbouring hydrogen nuclei. If an adjacent **carbon** atom has hydrogen atoms bonded to it, they will cause the peaks to split as shown in the table below.

1 neighbouring H atom	Peak splits into 2 lines (a doublet)
2 neighbouring H atoms	Peak splits into 3 lines (a triplet)
n neighbouring H atoms	Peak splits into (*n* + 1) lines

Ethanol has three peaks:

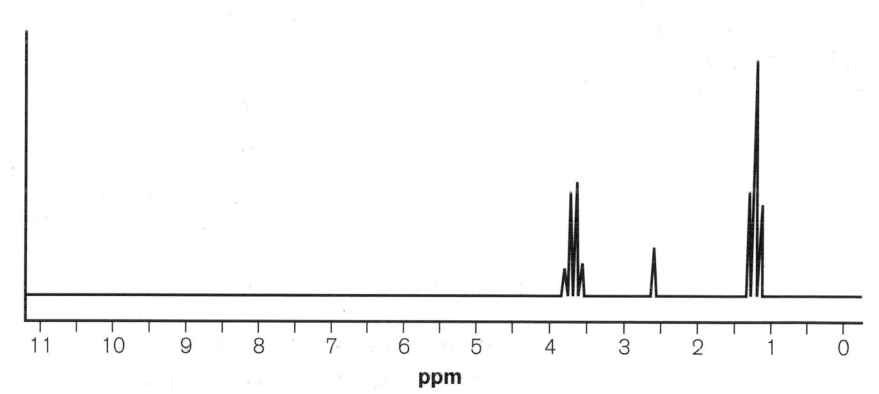

Figure 5.5 High resolution NMR spectrum of ethanol

In the high resolution NMR spectrum of ethanol, there is one peak due to the OH hydrogen, which is a single line (as it is hydrogen bonded); one peak due to the CH_2 hydrogen nuclei, which is split into four lines by the three H atoms on the neighbouring CH_3 group; 1 peak due to the CH_3 hydrogen nuclei, which is split into three lines by the two H atoms on the neighbouring CH_2 group.

Organic synthesis

★ You may be asked to suggest a synthetic route of up to four steps.

★ If the number of carbon atoms in the chain is *increased* by one carbon atom, consider a carbonyl compound reaction with HCN.

★ If the carbon chain is *decreased* by one carbon atom, consider the iodoform reaction.

Summary of possible synthetic routes

Starting from an alkene:

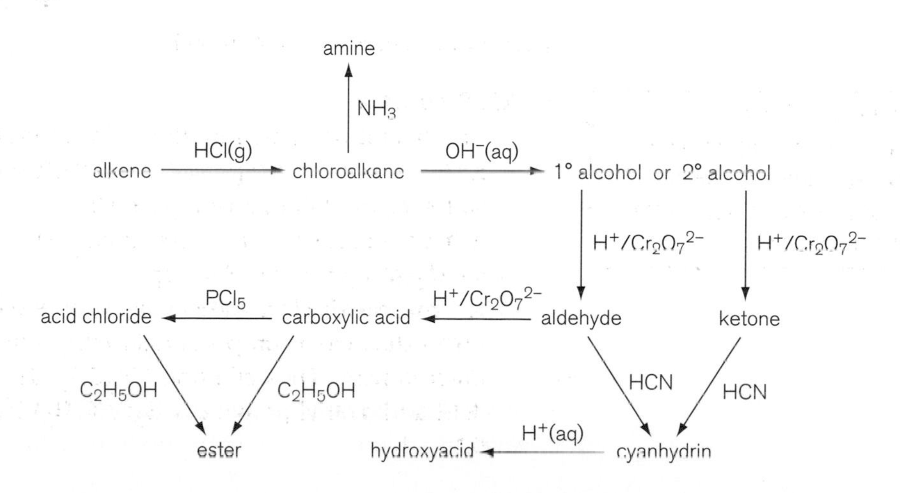

Starting from benzene:

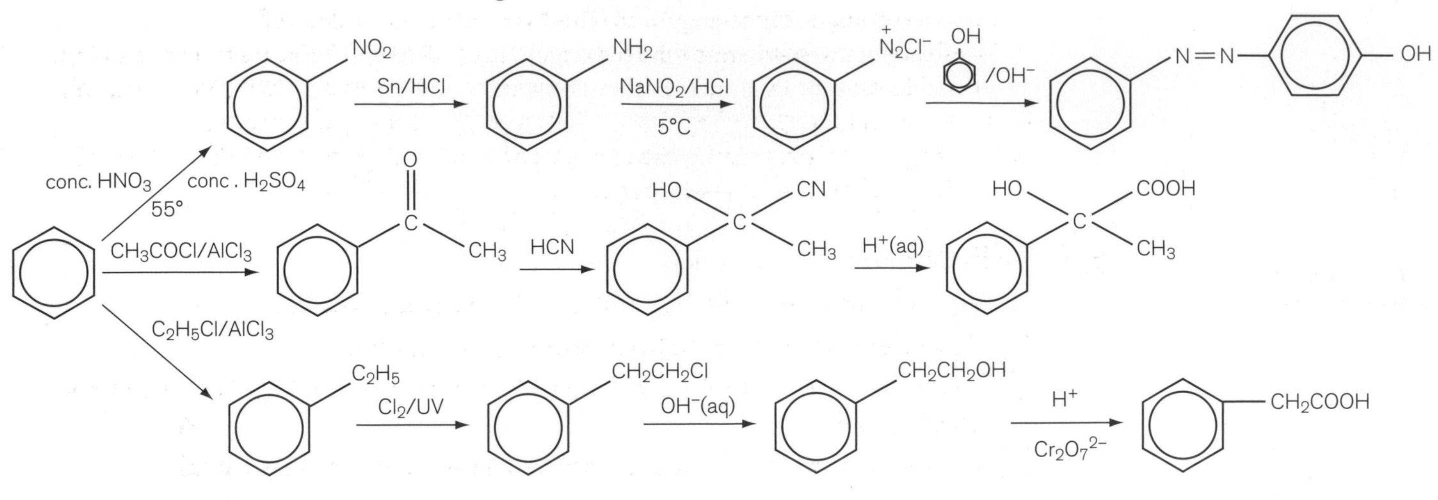

Hint

When a specific safety precaution is asked for, do not give the use of safety glasses, lab coats etc. as examples. The answer is probably either no naked flames (ether solvent) or wear gloves (harmful by skin absorption) or carry out the experiment in a fume cupboard, in which case the specific hazard, for example toxicity, must be stated.

The pure solid will have a sharp melting temperature.

Practical techniques: preparation

★ Heating under reflux is necessary when the reactant is volatile and the reaction is slow at room temperature.

★ Safety precaution: you must use a fume cupboard when a reactant or product is toxic, irritant or carcinogenic.

Practical techniques: purification

★ **Recrystallisation** (for solids):

(1) Dissolve the solid in a minimum of hot solvent.

(2) Filter the hot solution through a pre-heated funnel to remove insoluble impurities.

(3) Allow to cool, then filter under reduced pressure to remove soluble impurities

(4) Wash with a little cold solvent and allow to dry.

★ **Distillation** is used to separate a volatile liquid from a mixture, e.g. ethanal from a mixture of ethanol, acid and potassium dichromate(VI).

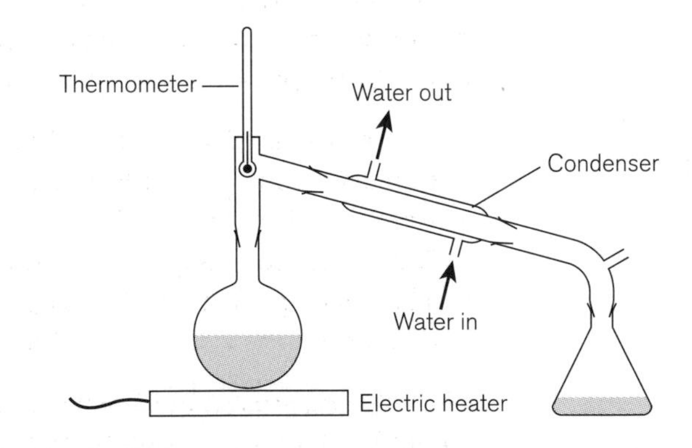

Figure 5.6 Apparatus for distillation

★ **Steam distillation** is used to separate a volatile substance from a mixture of solids and liquids, e.g. lemon oil from crushed lemon peel and water. Steam is blown into the mixture and the oil and water distil off, leaving behind the solids.

★ In **solvent extraction** an organic liquid, such as ether, which is immiscible with water, is added to an aqueous mixture of an organic liquid that is partially soluble

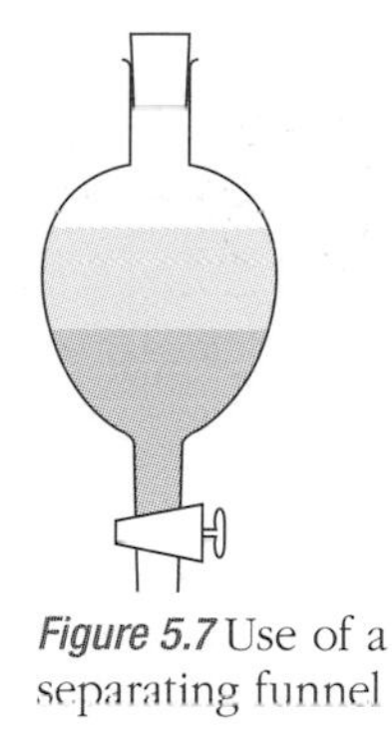

Figure 5.7 Use of a separating funnel

in water, e.g. phenylamine and water after steam distillation. The organic layer is removed using a separating funnel and the ether is distilled off.

★ **Washing with sodium carbonate** removes acid from a mixture of water and an insoluble organic compound. This is carried out by shaking with sodium carbonate solution in a separating funnel, followed by washing with water.

★ **Drying** — an organic liquid can be dried by standing it over lumps of solid calcium chloride or potassium hydroxide.

Checklist

Before attempting the questions on this topic, check that you can:

- ☐ deduce empirical formulae from percentage composition
- ☐ recall the tests for C=C, C–X (where X is a halogen), COH, C=O, CHO and COOH groups
- ☐ deduce the functional groups present from the results of chemical tests
- ☐ interpret infrared, NMR and mass spectra
- ☐ deduce pathways for the synthesis of organic molecules
- ☐ describe heating under reflux, recrystallisation, steam distillation and solvent extraction, and when they are used.

Testing your knowledge and understanding

★ What conclusion can be drawn about an organic compound that gives a red precipitate when mixed with Brady's reagent?

★ State the reagents and conditions needed for the conversion of:

a propan-1-ol to propanal

b benzene to nitrobenzene

c nitrobenzene to phenylamine

Answers

It is either an aldehyde or a ketone.

a Hot acidified potassium dichromate(VI); distil off the aldehyde as it is formed

b Concentrated sulfuric and nitric acids at 55°C

c Tin (or iron) and concentrated hydrochloric acid

The **answers** to the **numbered questions** are on page 146

1 Describe how would you distinguish between:

a pentan-3-one and pentanal

b 2-bromopropane and 2-chloropropane

c methylpropan-2-ol, methylpropan-1-ol and butan-2-ol

2 a A compound **X** contained 80.0% carbon, 6.7% hydrogen and 13.3% oxygen. Calculate its empirical formula.

b Compound **X** did not decolorise bromine water. It gave a red precipitate with Brady's reagent. Suggest the functional groups that are present in **X**.

c The mass spectrum of **X** had peaks at *m/e* values of 120, 105 and 77. Use these data and your answers to **a** and **b** to suggest a structural formula for **X**.

3 The infrared spectrum of aspirin showed peaks at 3200, 1720 and 1150 cm^{-1}. Use a data book to identify the groups that cause these peaks.

4 Low resolution NMR spectra of two aromatic isomers with molecular formula C_8H_{10} were obtained. The spectrum of isomer **Y** had three peaks of relative area 5:2:3; that of isomer **Z** had two peaks of relative area 6:4. Suggest structural formulae for **Y** and **Z**.

5 Outline how 2-hydroxypropanoic acid, $CH_3CH(OH)COOH$, can be synthesised from ethanol. Give all the reagents, the conditions and the intermediates.

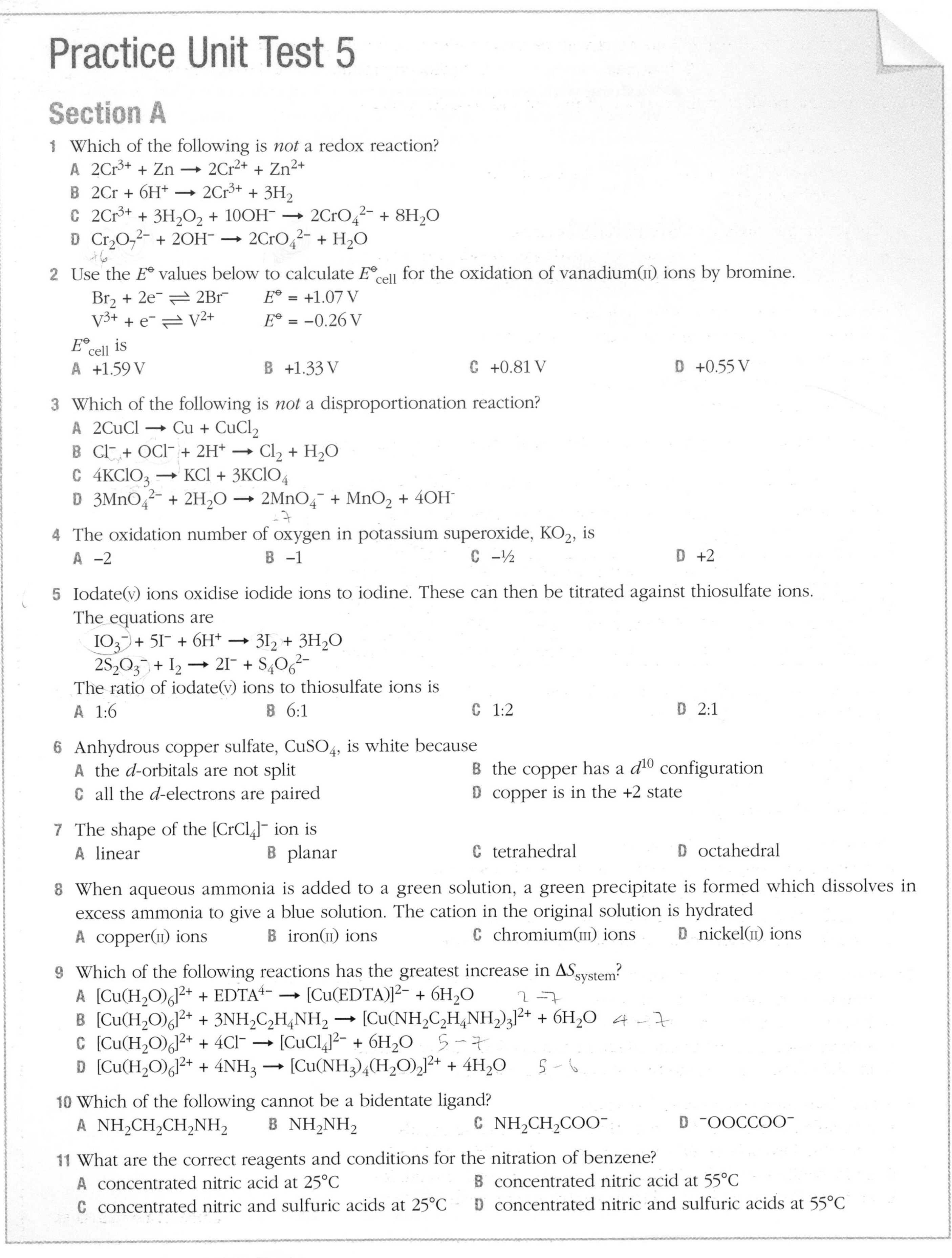

Practice Unit Test 5

Section A

1 Which of the following is *not* a redox reaction?

A $2Cr^{3+} + Zn \longrightarrow 2Cr^{2+} + Zn^{2+}$

B $2Cr + 6H^+ \longrightarrow 2Cr^{3+} + 3H_2$

C $2Cr^{3+} + 3H_2O_2 + 10OH^- \longrightarrow 2CrO_4^{2-} + 8H_2O$

D $Cr_2O_7^{2-} + 2OH^- \longrightarrow 2CrO_4^{2-} + H_2O$

2 Use the $E^\ominus$ values below to calculate $E^\ominus_{cell}$ for the oxidation of vanadium(II) ions by bromine.

$Br_2 + 2e^- \rightleftharpoons 2Br^-$ $\quad E^\ominus = +1.07\,V$

$V^{3+} + e^- \rightleftharpoons V^{2+}$ $\quad E^\ominus = -0.26\,V$

$E^\ominus_{cell}$ is

A +1.59 V **B** +1.33 V **C** +0.81 V **D** +0.55 V

3 Which of the following is *not* a disproportionation reaction?

A $2CuCl \longrightarrow Cu + CuCl_2$

B $Cl^- + OCl^- + 2H^+ \longrightarrow Cl_2 + H_2O$

C $4KClO_3 \longrightarrow KCl + 3KClO_4$

D $3MnO_4^{2-} + 2H_2O \longrightarrow 2MnO_4^- + MnO_2 + 4OH^-$

4 The oxidation number of oxygen in potassium superoxide, KO_2, is

A −2 **B** −1 **C** −½ **D** +2

5 Iodate(V) ions oxidise iodide ions to iodine. These can then be titrated against thiosulfate ions. The equations are

$IO_3^- + 5I^- + 6H^+ \longrightarrow 3I_2 + 3H_2O$

$2S_2O_3^- + I_2 \longrightarrow 2I^- + S_4O_6^{2-}$

The ratio of iodate(V) ions to thiosulfate ions is

A 1:6 **B** 6:1 **C** 1:2 **D** 2:1

6 Anhydrous copper sulfate, $CuSO_4$, is white because

A the *d*-orbitals are not split

B the copper has a d^{10} configuration

C all the *d*-electrons are paired

D copper is in the +2 state

7 The shape of the $[CrCl_4]^-$ ion is

A linear **B** planar **C** tetrahedral **D** octahedral

8 When aqueous ammonia is added to a green solution, a green precipitate is formed which dissolves in excess ammonia to give a blue solution. The cation in the original solution is hydrated

A copper(II) ions **B** iron(II) ions **C** chromium(III) ions **D** nickel(II) ions

9 Which of the following reactions has the greatest increase in ΔS_{system}?

A $[Cu(H_2O)_6]^{2+} + EDTA^{4-} \longrightarrow [Cu(EDTA)]^{2-} + 6H_2O$

B $[Cu(H_2O)_6]^{2+} + 3NH_2C_2H_4NH_2 \longrightarrow [Cu(NH_2C_2H_4NH_2)_3]^{2+} + 6H_2O$

C $[Cu(H_2O)_6]^{2+} + 4Cl^- \longrightarrow [CuCl_4]^{2-} + 6H_2O$

D $[Cu(H_2O)_6]^{2+} + 4NH_3 \longrightarrow [Cu(NH_3)_4(H_2O)_2]^{2+} + 4H_2O$

10 Which of the following cannot be a bidentate ligand?

A $NH_2CH_2CH_2NH_2$ **B** NH_2NH_2 **C** $NH_2CH_2COO^-$ **D** $^-OOCCOO^-$

11 What are the correct reagents and conditions for the nitration of benzene?

A concentrated nitric acid at 25°C

B concentrated nitric acid at 55°C

C concentrated nitric and sulfuric acids at 25°C

D concentrated nitric and sulfuric acids at 55°C

12 Which of the following will *not* react with benzene under appropriate conditions?
A hydrogen B bromine C hydrogen bromide D bromoethane

13 The organic product of the reaction of phenol and bromine water is
A 2-bromophenol
B 4-bromophenol
C a mixture of 2-bromophenol and 4-bromophenol
D 2,4,6-tribromophenol

14 Phenylamine is slightly soluble in water because
A it is polar
B it can form hydrogen bonds with water
C it is ionic
D it reacts with water to form an acid

15 Poly(ethenol) is soluble in water because
A it has an –OH group on every other carbon atom
B it is biodegradable
C it splits up into its monomers in water
D it is unstable in water

16 Ethyl ethanoate can be prepared from bromoethane in a four-step synthesis.

$$C_2H_5Cl \xrightarrow{\text{step 1}} C_2H_5OH \xrightarrow{\text{step2}} CH_3COOH \xrightarrow{\text{step 3}} CH_3COCl \xrightarrow{\text{step 4}} CH_3COOC_2H_5$$

What are the correct reagents and conditions for these four steps?

	Step 1	Step 2	Step 3	Step 4
A	KOH(aq)	$H^+/Cr_2O_7^{2-}$; heat under reflux	PCl_5	$C_2H_5OH(l)$
B	KOH(aq)	$H^+/Cr_2O_7^{2-}$; distil off as formed	PCl_3	$C_2H_5OH(aq)$
C	KOH in ethanol	$H^+/Cr_2O_7^{2-}$; heat under reflux	PCl_5	$C_2H_5OH(l)$
D	KOH in ethanol	$H^+/Cr_2O_7^{2-}$; distil off as formed	PCl_3	$C_2H_5OH(aq)$

17 An organic compound, **Q**, has the molecular formula $C_5H_{10}O$. Its mass spectrum has peaks at *m/e* of (M–15), (M–29), and (M–43). **Q** could be
A $(CH_3)_3CCHO$
B $(CH_3)_2CHCOCH_3$
C $CH_3CH_2CH_2COCH_3$
D $CH_3CH_2CH{=}CHCH_2OH$

18 The NMR spectrum of $HOCH_2CH{=}CHCH_2OH$ has
A 3 peaks: one singlet, one doublet and one split into four
B 3 peaks: one singlet, one triplet and one split into four
C 6 peaks: two singlets, two doublets and two triplets
D 6 peaks: two singlets, two triplets and two split into four

19 Impure benzoic acid can be recrystallised from hot water. The impurities removed are
A soluble impurities in both filtrations
B insoluble impurities in both filtrations
C soluble impurities in the first filtration and insoluble in the second
D insoluble impurities in the first filtration and soluble in the second

20 Steam distillation can be used to extract
A a volatile soluble liquid from a mixture of a solid and liquids
B a volatile insoluble liquid from a mixture of a solid and liquids
C an involatile soluble liquid from a mixture of a solid and liquids
D an involatile insoluble liquid from a mixture of a solid and liquids

Section A total: 20 marks

Section B

21 The half-equations needed for this question are:

$MnO_4^- + 8H^+ + 5e^- \rightleftharpoons Mn^{2+} + 4H_2O$ $\quad E^\ominus = +1.52\,V$

$2CO_2 + 2H^+ + 2e^- \rightleftharpoons (COOH)_2$ $\quad E^\ominus = -0.49\,V$

a **(i)** Describe how you would measure the standard reduction potential of the MnO_4^-, H^+/Mn^{2+} system. **(4 marks)**

(ii) Why is a reference electrode needed? **(1 mark)**

b **(i)** Calculate $E^\ominus_{cell}$ for the oxidation of ethanedioic acid, $(COOH)_2$, by manganate(VII) ions in acid solution and write the ionic equation for the reaction. **(3 marks)**

(ii) In practice this reaction does not take place unless it is heated. Suggest a reason for this. **(1 mark)**

(iii) What is the oxidation number of carbon in ethanedioic acid? **(1 mark)**

c This reaction can be used to estimate the concentration of a solution of ethanedioic acid, $(COOH)_2$. 25.0 cm^3 samples of ethanedioic acid solution were acidified and titrated against 0.0225 mol dm^{-3} potassium manganate(VII) solution. The mean titre was 26.75 cm^3. Calculate the concentration of the ethanedioic acid solution in g dm^{-3}. **(4 marks)**

Total: 14 marks

22 a Write the electron configuration of a Cu^+ and a Cu^{2+} ion. **(1 mark)**

b Explain why compounds of both Cu^{2+} and Cu^+ ions exist. **(2 marks)**

c Define:

(i) a transition metal **(1 mark)**

(ii) a bidentate ligand **(1 mark)**

(iii) the coordination number of a complex ion **(1 mark)**

d Explain why:

(i) $[Cu(H_2O)_6]^{2+}$ ions are coloured **(3 marks)**

(ii) $[CuCl_4]^{2-}$ ions are a different colour from that of the $[Cu(H_2O)]_6]^{2+}$ ion **(1 mark)**

(iii) anhydrous $CuSO_4$ is colourless **(2 marks)**

(iv) $[CuCl_2]^-$ ions are colourless **(1 mark)**

e Write the equation for the reaction of hydrated copper(II) ions with excess 1,2-diaminoethane, $NH_2CH_2CH_2NH_2$. **(1 mark)**

f Copper(I) chloride, CuCl, is insoluble in water but reacts with aqueous ammonia to give a colourless solution that slowly turns blue on exposure to air. Give the formulae of the species formed and explain the types of reaction occurring. **(4 marks)**

Total: 18 marks

23 a **(i)** Define the term 'electrophile'. **(2 marks)**

(ii) Write the mechanism for the reaction of benzene with chloroethane. Include an equation showing the formation of the electrophile. **(4 marks)**

b Nitrobenzene (density 1.2 g cm^{-3}, boiling point 211°C) can be prepared from benzene by the following method.

★ 10 cm^3 of benzene were carefully mixed with 20 cm^3 of a mixture of concentrated nitric and sulfuric acids and then warmed in a water bath at between 50°C and 60°C for 30 minutes.

★ 50 cm^3 of water were added carefully and the organic layer removed using a separating funnel.

★ The organic layer was washed with sodium carbonate solution and then with water.

★ Lumps of anhydrous calcium chloride were added to the organic layer and left for 15 minutes.

★ The organic layer was then distilled. The fraction that boiled over between 198° and 214°C was collected.

(i) Write the equation for the reaction between benzene and the acid mixture. **(1 mark)**

(ii) Explain why the temperature must neither drop below 50°C nor rise above 60°C **(2 marks)**

(iii) Explain which layer in the separating funnel contained the nitrobenzene. **(1 mark)**

(iv) Why was the organic layer washed with sodium carbonate solution? **(1 mark)**

(v) What is the purpose of the calcium chloride? **(1 mark)**

c (i) State the reagents used to prepare phenylamine from nitrobenzene. **(2 marks)**

(ii) Write equations for the reaction of phenylamine with:
- H^+ ions
- ethanoyl chloride, CH_3COCl **(2 marks)**

(iii) Phenylamine can be converted to a solution containing diazonium ions by reaction with sodium nitrite, $NaNO_2$, and acid. Write the *ionic* equation for this reaction. **(1 mark)**

(iv) The diazonium ion reacts with phenol to form a yellow dye. Write the structural formula of the yellow dye. **(1 mark)**

Total: 18 marks

Section B total: 50 marks

Section C

24 Read the passage below and then answer the questions that follow.

Forensic chemistry

Forensic chemistry depends on accurate identification of substances and the amounts of substances present in tissues. The first stage is often the separation of mixtures using chromatographic techniques such as HPLC and GLC. Identification may proceed using combustion analysis to determine the percentage of each element present. Chemical tests are then used to identify the functional groups present. Finally, the unique compound must be identified. This is achieved by, for example, preparing a solid derivative and measuring its melting point or by spectroscopic analysis.

NMR spectroscopy is a useful method because the number of peaks and the splitting pattern can be used to identify a particular isomer. The number of different chemical environments for the hydrogen nuclei determine the number of peaks and the splitting pattern is determined by the $(n + 1)$ rule, where n is the number of hydrogen atoms on adjacent carbon atoms. Thus, the symmetrical butane-1,4-dioic acid, $HOOCCH_2CH_2COOH$, has two peaks, one a singlet (COO**H**) and the other (**CH$_2$**) a triplet.

The dentine in teeth contains the L- enantiomer of aspartic acid, $NH_2CH(CH_2COOH)COOH$. As the person grows older, the L-aspartic acid changes to the D-form and this process continues after death. This means that skulls can be dated by measuring the ratio of D-aspartic acid to L-aspartic acid in the teeth. This is one important way of dating skulls found in archeological sites and it is effective in skulls up to 1500 years old.

a (i) An organic compound **Z** was thought to be the main ingredient in a perfume. It was found to contain 47.06% carbon, 5.88% hydrogen and 47.06% oxygen. Use these data to show that the empirical formula is $C_4H_6O_3$. **(2 marks)**

(ii) Compound **Z** was tested as below. State what can be deduced from the results of each test.
- When phosphorus pentachloride was added there were clouds of an acidic gas. **(1 mark)**
- When **Z** was added to a solution of sodium carbonate, bubbles of gas were produced which turned limewater cloudy. **(1 mark)**
- When Brady's reagent was added, there was a yellow precipitate. **(1 mark)**
- When warmed with Tollens' reagent, a silver mirror was produced. **(1 mark)**

(iii) 2.36 g of compound **Z** was dissolved in water and the solution made up to 250 cm^3 with distilled water. 25.0 cm^3 portions of this solution were titrated against 0.105 mol dm^{-3} sodium hydroxide solution. The mean titre was 22.05 cm^3. Calculate the molar mass of **Z** and, using your answer to **(i)**, calculate its molecular formula. **(4 marks)**

(iv) Write the structural formula of *two* substances that could be **Z**. **(2 marks)**

(v) The NMR spectrum of **Z** had four lines, one singlet, two triplets and one peak split into four. Identify which of the two isomers you drew in (iv) is **Z**. Justify your answer. **(2 marks)**

b Aspartic acid is a chiral molecule found in the dentine of teeth.

(i) Draw the structures of the two enantiomers of aspartic acid and say how they could be distinguished from each other in the laboratory. **(3 marks)**

(ii) The change from L-aspartic acid to D-aspartic acid is a first-order reaction with a half-life of approximately 250 years. Calculate the age of a skull that was found to have a ratio of L-aspartic acid to D-aspartic acid of 1:3. **(2 marks)**

(iii) If the error in the value of the half-life is ±50 years, calculate the error in dating the skull. **(1 mark)**

Total: 20 marks

Paper total: 90 marks

Unit 1 The core principles of chemistry

Formulae, equations and moles

1 a

Element	%	Divide by smallest
C	82.76/12.0 = 6.90	1
H	17.24/1.0 = 17.24	2.5

Therefore, empirical formula is C_2H_5.

b mass of C_2H_5 = 29.0, but M_r = 58.0
Therefore, molecular formula is C_4H_{10}.

2 a $4NH_3 + 5O_2 \longrightarrow 4NO + 6H_2O$
b $2Fe^{3+}(aq) + Sn^{2+}(aq) \longrightarrow 2Fe^{2+}(aq) + Sn^{4+}(aq)$

Hint
Ionic equations must also balance for charge. Here, both sides are 8+.

3 equation: $H_3PO_4 + 3NaOH \longrightarrow Na_3PO_4 + 3H_2O$
amount of H_3PO_4 = 2.34/98.0 = 0.02388 mol
amount of NaOH = 0.02388 × 3/1 = 0.07164 mol
mass of NaOH = 0.07164 × 40.0 = 2.87 g

4 a atom economy $= \dfrac{\text{molar mass of } PbCl_2 \times 100}{\text{molar masses of reactants}}$
$= \dfrac{278.2 \times 100}{2 \times 58.5 + 331.2} = 62.1\%$

b It would be higher. The molar mass of HCl is less than that of NaCl, so the denominator would be smaller and, therefore, the atom economy greater.

5 amount of $Bi(NO_3)_3 = 0.025\,dm^3 \times 0.55\,mol\,dm^{-3}$
$= 0.01375$ mol
amount of H_2S = 0.01375 × 3/2 = 0.0206 mol
volume of H_2S gas = mol × molar volume
= 0.0206 × 24 = 0.49 dm^3

6 mol of $H_2(g)$:mol of $CH_4(g)$ = 3:1
volume of $H_2(g)$:volume of $CH_4(g)$ = 3:1
volume of $H_2(g)$ = 3 × 33 = 99 dm^3

7 a 0.030 g of $Al_2(SO_4)_3$ in $10^6\,cm^3$ water,
so 0.030×10^{-6} g in 1 cm^3 of water
$\dfrac{0.030 \times 10^{6}}{342.3} = 8.76 \times 10^{-11}$ mol of $Al_2(SO_4)_3$
$= 2 \times 8.76 \times 10^{-11}$ mol of Al^{3+} ions
number of Al^{3+} ions is $2 \times 8.76 \times 10^{-11} \times 6.02 \times 10^{23}$
$= 1.06 \times 10^{14}$

b 0.030×10^{-9} mol of $Al_2(SO_4)_3$ in 1 dm^3,
so 3.0×10^{-14} moles in 1 cm^3
so, there are $2 \times 3.0 \times 10^{-14}$ moles of Al^{3+} in 1 cm^3
and $6.0 \times 10^{-14} \times 6.02 \times 10^{23} = 3.6 \times 10^{10}$ Al^{3+} ions in 1 cm^3

8 4.0 g of NaOH = 4.0/40.0 = 0.100 mol, which would make 0.050 mol Na_2SO_4
5.0 g of H_2SO_4 = 5.0/98.1 = 0.051 mol, which would make 0.051 mol Na_2SO_4
Therefore, H_2SO_4 is in excess, as all the NaOH reacts.
0.50 mol Na_2SO_4 = 0.50 × 122.1 = 6.1 g Na_2SO_4 produced.

Energetics

1 A pressure of 1 atmosphere, a stated temperature (298K) and all solutions of concentration 1.00 mol dm^{-3}. Each substances in its most stable state.

2 a $2C(s) + 2H_2(g) + O_2(g) \longrightarrow CH_3COOH\ (l)$
b $CH_3COOH(l) + 2O_2(g) \longrightarrow 2CO_2(g) + 2H_2O(l)$
c $CH_3COOH(aq) + NaOH(aq) \longrightarrow CH_3COONa(aq) + H_2O(l)$
or $CH_3COOH(aq) + OH^-(aq) \longrightarrow CH_3COO^-(aq) + H_2O(l)$

Hint
Note that ethanoic acid is a weak acid, and so its formula must be written in full and not just as H^+.

3

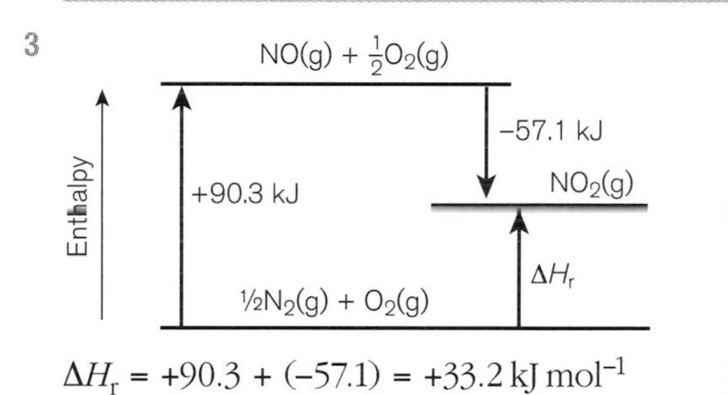

$\Delta H_r = +90.3 + (-57.1) = +33.2\,kJ\,mol^{-1}$

Hint
Note that the + sign is given in the answer in order to emphasise that the reaction is endothermic.

4 $\Delta H_r = \Delta H_f\,(N_2O_4) - 2 \times \Delta H_f\,(NO_2)$
$= +9.7 - 2 \times (33.9) = -58.1\,kJ\,mol^{-1}$

5

$12C(s) + 12H_2(g) + O_2(g) \xrightarrow{\Delta H_f} CH_3(CH_2)_{10}COOH(s)$
ΔH_1
ΔH_2
$12CO_2(g) + 12H_2O(l)$

$\Delta H_f = \Delta H_1 + \Delta H_2$
where $\Delta H_1 = 12 \times \Delta H_c$(C(graphite)) + 12 × ΔH_c(hydrogen)
and $\Delta H_2 = -\,\Delta H_c$(lauric acid)
$\Delta H_f = (12 \times (-394)) + (12 \times (-286)) - (-7377)$
$= -783\,kJ\,mol^{-1}$

6 $HCl + NaOH \longrightarrow NaCl + H_2O$

amount of HCl = $V(dm^3) \times$ concentration ($mol\,dm^{-3}$)

$$= \frac{100}{1000} \times 1.00 = 0.100\,mol$$

rise in temperature = +6.80°C

heat produced when 0.100 mol reacts = total mass × specific heat capacity × rise in temperature

$= 200\,g \times 4.18\,J\,g^{-1}{}^{\circ}C^{-1} \times 6.80^{\circ}C = 5685\,J = 5.685\,kJ$

heat produced when 1 mol reacts = 5.6848/0.100

$= 56.8\,kJ\,mol^{-1}$

$\Delta H_{neut} = -56.8\,kJ\,mol^{-1}$

Hint

The mixture became hotter, so the reaction is exothermic. The mass used is the mass of the whole solution:

$100\,cm^3 + 100\,cm^3 = 200\,cm^3 = 200\,g$

It is not the mass of the HCl.

7 The Hess's law diagram is

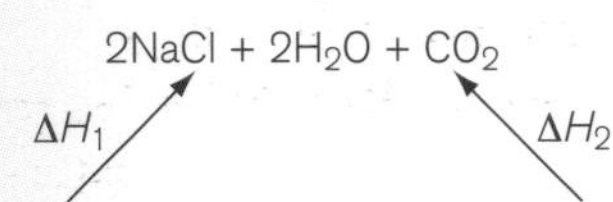

where $\Delta H_1 = 2 \times$ enthalpy of the reaction of $NaHCO_3$ with acid

and where ΔH_2 = enthalpy of reaction of Na_2CO_3 with acid

Method: take 0.02 mol of solid $NaHCO_3$ and add $20\,cm^3$ of $1.0\,mol\,dm^{-3}$ HCl and measure the temperature rise.

Take 0.01 mol of solid Na_2CO_3 and add $20\,cm^3$ of $1.0\,mol\,dm^{-3}$ HCl and measure the temperature rise.

Calculate heat change as $20 \times 4.18 \times \Delta T$ for each, then calculate ΔH_1 as –heat change/0.02 and ΔH_2 as –heat change/0.01.

Enthalpy of decomposition of $NaHCO_3 = 2 \times \Delta H_1 - \Delta H_2$.

8

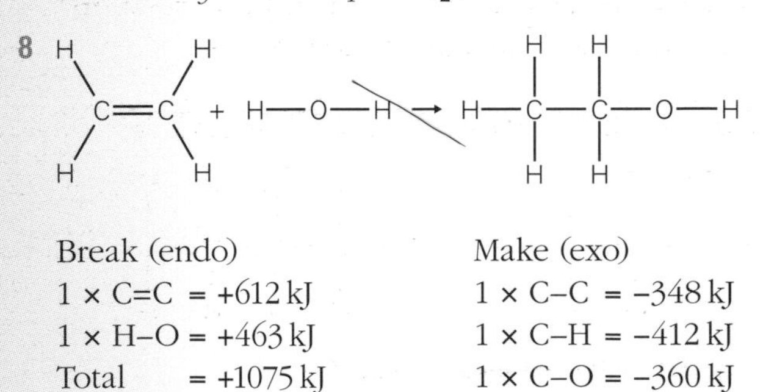

Break (endo)		Make (exo)	
1 × C=C	= +612 kJ	1 × C–C	= –348 kJ
1 × H–O	= +463 kJ	1 × C–H	= –412 kJ
Total	= +1075 kJ	1 × C–O	= –360 kJ
		Total	= –1120 kJ

$\Delta H = +1075 + (-1120) = -45\,kJ\,mol^{-1}$

Atomic structure and the periodic table

1 The relative **isotopic** mass is the mass of a single isotope of an element (relative to 1/12 the mass of a carbon-12 atom). The relative **atomic** mass is the average mass taking into account the isotopes of the element. The relative isotopic masses of the two chlorine isotopes are 35 and 37, but the relative atomic mass of chlorine is 35.5.

2 There are many more ^7Li atoms than ^6Li atoms in natural lithium, so the average is close to 7.

3 A_r of magnesium =

$$\frac{((24 \times 78.6) + (25 \times 10.1) + (26 \times 11.3))}{100} = 24.3$$

4 Let the % of ^{69}Ga be x, and so the % of ^{71}Ga $= (100 - x)$

$$69.8 = \frac{[69.0x + 71.0(100 - x)]}{100}$$

$\therefore 6980 = 69x + 7100 - 71x$

$\therefore 2x = 120$

$\therefore x = 60$

Gallium contains 60% of the ^{69}Ga isotope.

5

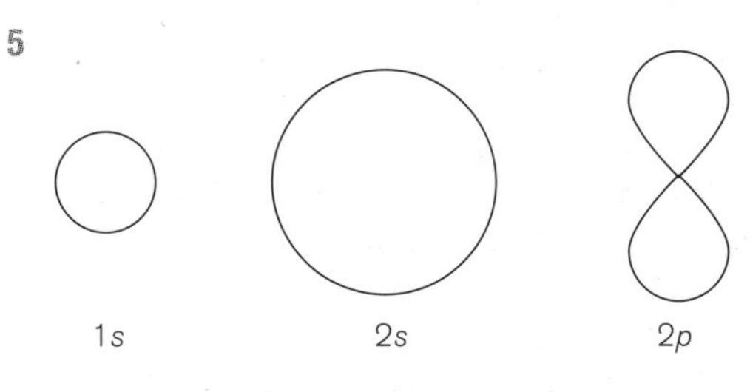

6 Group 5. There is a big jump between the fifth and sixth ionisation energies. Therefore, the sixth electron is being removed from a lower shell (nearer to the nucleus) than the first five electrons.

7 a

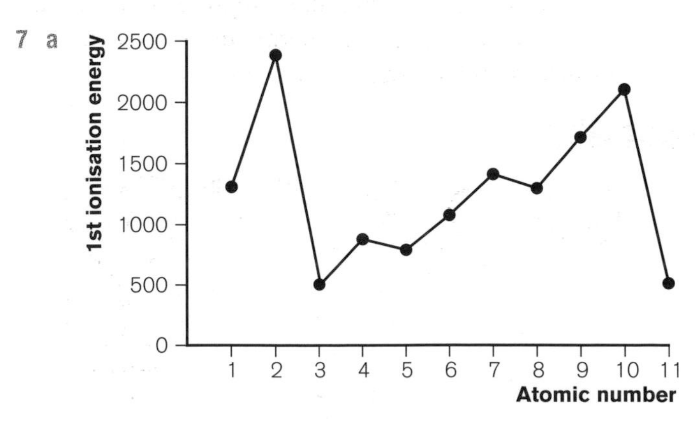

b (i) A $2p$ electron, which is in a higher energy level, is being removed. So less energy is required to remove it compared with that required to remove a $2s$ electron.
(ii) There are two electrons in the $2p_x$-orbital that are repelling one another, so less energy is required to remove one of the pair than if there had been only one electron in that orbital.
(iii) A $3s$ electron is further from the nucleus than a $2s$ electron, so less energy is required to remove it.

8 The electron in a box notation for chlorine is:

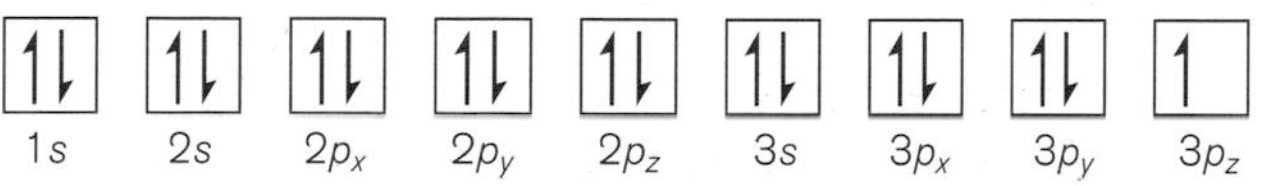

9 All have one electron in their outer shells, but K is larger (larger atomic radius) than Na, which is larger than Li. In the larger atoms, the outer electron is held less firmly (the increased nuclear charge is balanced by the increased number of shielding electrons). It is easiest to remove an electron from potassium and hardest to remove one from Li, This makes potassium the most reactive.

10 a (i) $Li(g) + e^- \rightarrow Li^-(g)$
(ii) $Cl(g) + e^- \rightarrow Cl^-(g)$
(iii) $O(g) + e^- \rightarrow O^-(g)$

Hint

Note that the species on the left-hand side is a gaseous *atom* and the species on the right is a *negative* ion.

b $O^-(g) + e^- \rightarrow O^{2-}(g)$

c For the first electron, the negatively charged electron is brought closer to the positive nucleus in a *neutral* atom and so energy is released and the electron affinity is exothermic. For the second electron affinity, the negative electron is brought towards a *negative* ion and so repulsion has to be overcome. This requires energy and the electron affinity is endothermic.

Bonding

1 a

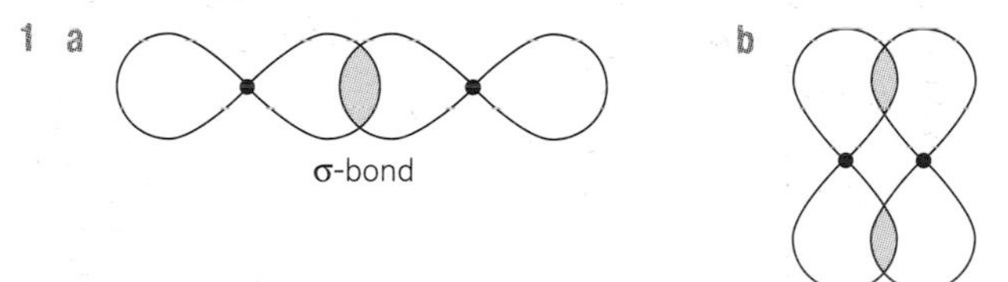

σ-bond

b

π-bond

2 An ionic bond is the electrostatic attraction between a positive and a negative ion.

3 a Isoelectronic means that they have identical electron configurations.

b $Ca^{2+} < K^+ < Cl^- < S^{2-}$

4 a

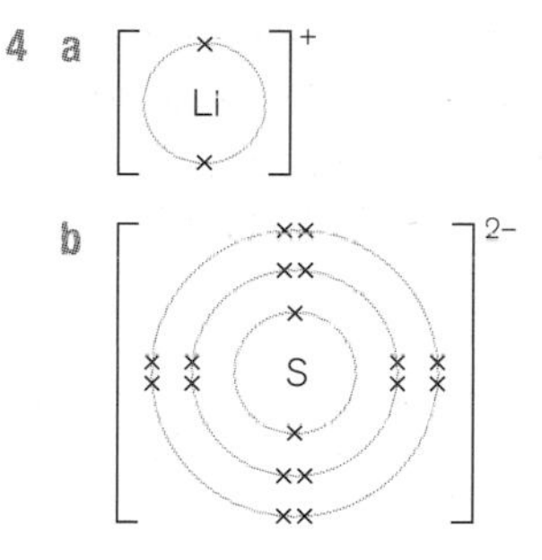

b

5 a K(g)

b Step 1 — enthalpy of atomisation of potassium;
step 2 — first ionisation energy of potassium;
step 3 — enthalpy of atomisation of fluorine;
step 4 — electron affinity of fluorine.

c ΔH_f (KF) = −563 = +90 + 418 + 79.1 + (−348) + lattice energy

Lattice energy = −563 − 90 − 418 − 79.1 +348 = −802.1 kJ mol^{-1}

6 A covalent bond is the attraction of the nuclei of two atoms for a shared pair of electrons.

7 a :F:O:F: (dot-and-cross)

b O::O (dot-and-cross)

8 A metallic bond is the force of attraction between the delocalised electrons and the metal ions.

9 In Na(s) there are loosely held delocalised electrons that can move through the lattice. In NaCl(s) the electrons are localised on the ions and so not loosely held *and* the ions are fixed in their positions in the lattice and, therefore, are not free to move.

Introductory organic chemistry

1

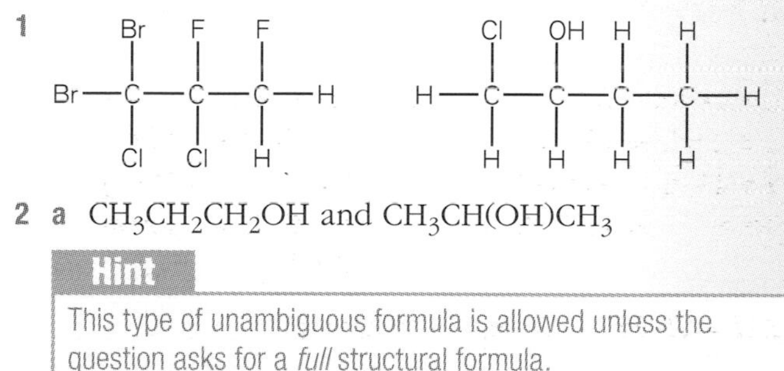

2 a $CH_3CH_2CH_2OH$ and $CH_3CH(OH)CH_3$

Hint

This type of unambiguous formula is allowed unless the question asks for a *full* structural formula.

b $CH_3CH_2CH_2CH_2CH_3$ $CH_3C(CH_3)_2CH_3$ $CH_3CH_2CH(CH_3)CH_3$

c $CH_3CH{=}CHCH_3$ $(CH_3)_2C{=}CH_2$ $CH_3CH_2CH{=}CH_2$.

3 *Z*-3-chloro-but-2-enoic acid

4 a A free radical is a species with a single unpaired electron, e.g. Cl•.

b Homolytic fission is when a bond breaks and one electron goes to each atom.

c An electrophile is a species that seeks out negative centres and accepts a pair of electrons to form a new covalent bond.

5 a $C_2H_6 + Cl_2 \longrightarrow C_2H_5Cl + HCl$
$C_2H_5Cl + Cl_2 \longrightarrow C_2H_4Cl + HCl$ and so on
Ultraviolet light

b $C_2H_6 + 3\frac{1}{2}O_2 \longrightarrow 2CO_2 + 3H_2O$
A flame or a spark

6 a $CH_3CH{=}CH_2 + H_2 \longrightarrow CH_3CH_2CH_3$
Heat and Pt/Ni catalyst

b $CH_3CH{=}CH_2 + Br_2 \longrightarrow CH_3CHBrCH_2Br$
Solution in hexane
Goes from red-brown to colourless

c $CH_3CH{=}CH_2 + HI \longrightarrow CH_3CHICH_3$
Mix gases at room temperature

d $CH_3CH{=}CH_2 + [O] + H_2O \longrightarrow CH_3CH(OH)CH_2OH$
Aqueous
Goes from a purple solution to a brown precipitate

Practice Test Unit 1

Section A

1 C	**5** A	**9** C	**13** C	**17** B
2 B	**6** B	**10** D	**14** A	**18** C
3 D	**7** B	**11** D	**15** B	**19** B
4 A	**8** B	**12** C	**16** A	**20** D

Section A total: 20 marks

Section B

21 a It is the enthalpy change when 1 mol of a substance ✓ is formed from its elements in their most stable states ✓ at 1 atm pressure and a stated (usually 298 K) temperature ✓.

b $\Delta H_f = (2 \times \Delta H_c(C)) + (3 \times \Delta H_c(H_2)) - (\Delta H_c(C_2H_6))$
$= (2 \times -394) + (3 \times -286) - (-1560)$
$= -86\ kJ\ mol^{-1}$
use of ×2 ✓; use of ×3 ✓; answer plus unit ✓

c heat released $= 100 \times 4.18 \times 19 = 7942\ J$ ✓
amount of ethane $= 135/24\,000 = 0.005625\ mol$ ✓
$\Delta H_c = -7925/0.005625$ ✓
$= -1\,412\,000\ J\ mol^{-1}$ or $= -1.4 \times 10^3\ kJ\ mol^{-1}$
sign ✓ value and unit ✓

Hint

A sign, unit or phrase in brackets means that it is not necessary to score the mark. In c the third mark is for dividing the heat change by the moles. The sign of ΔH_c is negative as heat was given out and it is scored in the final answer.

d

Bonds broken		Bonds made	
C=C	+612	C–C	–348
H–H	+436	2 × C–H	2 × –412
Total	(+)1048 ✓	Total	(–)1172 ✓

$\Delta H_r = +1048 - 1172 = -124\ (kJ\ mol^{-1})$ sign ✓ value ✓

Total: 15 marks

22 a Put a spot of potassium manganate(VII) solution on a strip of filter paper soaked in sodium chloride solution ✓. Connect the ends of the paper to a supply of electricity ✓. Note that the purple colour (of MnO_4^- ions) moves towards the positive electrode ✓.

b (i) $[Ca]^{2+}$ with 8 electrons (××) shown around Ca ✓

(ii) $[Cl]^{-}$ with 8 electrons (××) shown around Cl ✓

c (i) $\Delta H_f = -795$
$= \Delta H_a$(calcium) + sum of 1st and 2nd IE(calcium) + (2 × ΔH_a(chlorine)) + (2 × EA(chlorine)) + LE
$= +193 + 590 + 1150 + (2 \times 121) + (2 \times (-364)) + LE$
$LE = -795 - 193 - 590 - 1150 - 242 + 728$
$= -2242\ kJ\ mol^{-1}$
Correct algebra ✓; use of 2× (twice) ✓; sign of answer ✓; value and unit ✓

(ii) The formation of $CaCl_3$ would require the third ionisation energy of calcium ✓. This is very endothermic because the third electron would have to be removed from the inner shell, which is poorly shielded from the nucleus as there are only ten electrons ✓ (shielding the +20 nucleus). This extra endothermic step will not be compensated for by the more exothermic lattice energy of $CaCl_3$ ✓, so the compound will not exist.

Hint

Do not give as your reason anything about the so-called 'stability of a noble gas structure'.

(iii) The real or experimental lattice energy was calculated in part (i) to be $-2242\,kJ\,mol^{-1}$. This is more exothermic than the theoretical value because the Ca^{2+} ion polarises ✓ the chloride ion leading to the bond being partially covalent ✓.

(iv) The bromide ion is larger than the chloride ion (and has the same charge of −1) ✓, so it is more easily polarised ✓ resulting in a greater difference between the Born–Haber and the theoretical lattice energies.

d It is the attraction between the shared pair of electrons and the nuclei of the two atoms ✓.

e (i) 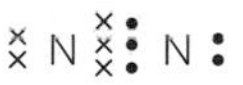 ×× N ×:×:×: N : ✓

(ii)

```
       ••
      : Cl :
  ••   ו   ••
: Cl × Si × Cl :
  ••   ו   ••
      : Cl :
       ••
```
✓

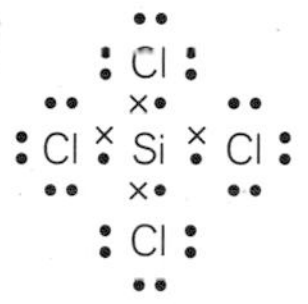

Total: 19 marks

23 (a) (i) A homologous series is a series of compounds with the same functional group ✓ and which have the same general formula/each member differs by CH_2 from the next ✓.

Hint

A forward slash represents an alternative scoring point.

(ii) 10 g octane = 10/114.0 = 0.08772 mol ✓
amount O_2 needed = 12.5 × 0.08772 = 1.096 mol ✓
volume of O_2 = moles × molar volume
= 1.096 × 24✓ = 26.3 dm³ ✓
(value and unit needed for the mark)

(b) (i) $CH_4 + Cl_2 \rightarrow CH_3Cl + HCl$ ✓

(ii) Initiation:

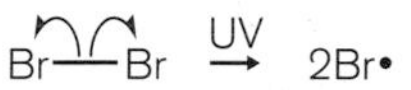

$Br{-}Br \xrightarrow{UV} 2Br\bullet$ ✓

Propagation: $Cl\bullet + CH_4 \rightarrow HCl + CH_3\bullet$ ✓
then $CH_3\bullet + Cl_2 \rightarrow CH_3Cl + Cl\bullet$ ✓
Termination: any one of $Cl\bullet + Cl\bullet \rightarrow Cl_2$ or $CH_3\bullet + CH_3\bullet \rightarrow C_2H_6$ or $Cl\bullet + CH_3\bullet \rightarrow CH_3Cl$ ✓

(c) (i) Add bromine water ✓. Goes from (red-)brown to colourless ✓

(ii) $CH_2{=}CHCH_3 + Br_2 \rightarrow CH_2BrCHBrCH_3$ ✓

(iii) 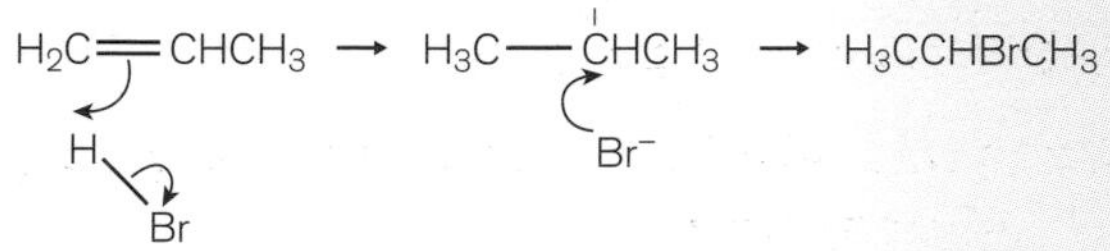 $H_2C{=}CHCH_3$ (with H–Br) $\rightarrow H_3C{-}\overset{+}{C}HCH_3$ (Br^-) $\rightarrow H_3CCHBrCH_3$

both arrows ✓ intermediate ✓ arrows from Br ✓

(iv) The secondary carbocation $CH_3C^+HCH_3$ intermediate ✓ is more stable than the primary carbocation, which would lead to 1-bromopropane ✓

Total: 19 marks

24 a Across the period the number of protons in the nucleus increases ✓, but the number of shielding/inner electrons remains the same ✓. Thus the attraction of the nucleus for the outer electron increases ✓.

b Magnesium is [Ne] $2s^2$ and aluminium is [Ne] $2s^2\,2p^1$. The $2p$ electron is in a higher energy level and so is easier to remove ✓.

c The number of shielding electrons increases by eight after each noble gas (whereas the number of protons only increases by one) and so the amount of shielding increases considerably ✓.

d Both the number of protons and the number of shielding electrons increase by eight ✓. The outer electron in potassium is further from the nucleus than that of sodium, so it is easier to remove ✓.

Total: 7 marks

Section B total: 60 marks

Paper total: 80 marks

About 56 marks would be needed for an A grade and about 27 for an E grade.

Unit 2 Application of core principles of chemistry

Shapes of molecules and ions, polarity and intermolecular forces

1

	Number of σ-bond pairs	Number of lone pairs	Total number of pairs	Shape
SiH_4	4	0	4	Tetrahedral
BF_3	3	0	3	Triangular planar
HC≡CH	2 (on each carbon)	0	2	Linear
PCl_3	3	1	4	Triangular pyramidal
H_2S	2	2	4	Bent
NH_4^+	4	0	4	Tetrahedral
PCl_6^-	6	0	6	Octahedral

2 Both contain polar bonds, but in CCl_4 the dipoles cancel because the molecule is symmetrical. In $CHCl_3$, the dipoles do not cancel, so the molecule has a dipole moment (is polar).

3 There is a large difference in electronegativity between fluorine and hydrogen and both are small atoms. The lone pair of electrons on the δ– F atom in one molecule of HF bonds onto the δ+ hydrogen atom in another HF molecule.

4

	Hydrogen bonds	Dispersion forces	Dipole–dipole forces
HF	**Yes**	**Yes**	**Yes**
I_2	No	**Yes**	No
HBr	No	**Yes**	**Yes**
PH_3	No	**Yes**	**Yes**
Ar	No	**Yes**	No

Hint
The strongest intermolecular force in all five molecules is the dispersion force.

5 **a** A molecule of ethane has fewer electrons than a molecule of propane, so the dispersion forces are weaker. Therefore, less energy is needed to separate the molecules and ethanol has the lower boiling point.

Hint
Dispersion forces are also called London, van der Waals or induced dipole forces.

b Propene has only two fewer electrons than propane, but more significantly the geometry around the double bond means that alkenes pack less well than alkanes.

c Branched-chain alkanes pack less well than unbranched alkanes, so methyl propane has the lower boiling point.

Hint
When answering a question about the boiling point of covalent molecular substances, *never* mention the strength of covalent bonds. Covalent bonds are *not* broken on boiling.

6 **a** NH_3. There are stronger *inter*molecular forces in NH_3 as it forms hydrogen bonds whereas PH_3 does not.

b HBr. A molecule of HBr has more electrons than a molecule of HCl and so it has stronger dispersion forces (which outweigh the difference in dipole–dipole forces).

c Propanone. Both substances have similar strength dispersion forces because they have the same number of electrons. However, propanone is polar so also has dipole–dipole forces.

d S_8. It has about twice as many electrons as P_4 and so has stronger dispersion forces.

e NaCl. It has ion–ion forces that are very much stronger than the weak intermolecular dispersion forces between CCl_4 molecules.

7 The chlorine atom in CH_3Cl is δ– but is too big to form hydrogen bonds with the δ+ hydrogen atoms in water. The δ– oxygen in CH_3OH uses its lone pairs of electrons to form hydrogen bonds with the δ+ hydrogen atoms in water; the δ+ hydrogen of the OH groups in CH_3OH form hydrogen bonds with the lone pairs on the δ– oxygen atoms in water.

Redox

1 **a** **(i)** $Sn^{2+}(aq) \rightarrow Sn^{4+}(aq) + 2e^-$

Hint
Electrons are on the right as Sn^{2+} is oxidised: $2e^-$ needed as oxidation number changes by 2

(ii) $Fe^{3+}(aq) + e^- \rightarrow Fe^{2+}(aq)$

b $Sn^{2+}(aq) + 2Fe^{3+}(aq) \rightarrow Sn^{4+}(aq) + 2Fe^{2+}(aq)$

Hint
Equation (ii) has to be doubled, so that the electrons cancel when two times equation (ii) is added to equation(i).

2 a (i) $H_2O_2(aq) + 2H^+(aq) + 2e^- \rightarrow 2H_2O(l)$

Hint

H_2O_2 is reduced, so the electrons are on the left.

(ii) $S(s) + 2H^+(aq) + 2e^- \rightarrow H_2S(aq)$

Hint

Two electrons are needed on the left since S is reduced and its oxidation number changes from 0 to −2.

b $H_2O_2(aq) + H_2S(aq) \rightarrow 2H_2O(l) + S(s)$

Hint

Both H_2S and H_2O_2 must be on the left as they are the reactants; equation (ii) is reversed and added to equation(i). Note that the H^+ ions cancel out.

3 a (i) $PbO_2(s) + 2H_2SO_4(aq) + 2e^- \rightarrow PbSO_4(s) + SO_4^{2-}(aq) + 2H_2O(l)$

Hint

$2e^-$ are needed on the left as PbO_2 is reduced and its oxidation number changes from +4 to +2. $PbSO_4$ is insoluble and so is precipitated.

(ii) $PbSO_4(s) + 2e^- \rightarrow Pb(s) + SO_4^{2-}(aq)$

b $PbO_2(s) + Pb(s) + 2H_2SO_4(aq) \rightarrow 2PbSO_4(s) + 2H_2O(l)$

Hint

$2e^-$ are needed on the left since $PbSO_4$ is reduced and the oxidation number of lead changes from +2 to 0. As Pb is a reactant, equation (ii) has to be reversed and then added to equation (i).

c It is not disproportionation. Although lead is being oxidised and reduced, the lead atoms on the left of the equation are not in the same species.

The periodic table: group 2

1 Dip a hot platinum (or nichrome) wire in clean concentrated HCl and then into each substance in turn, cleaning the wire between each test. The substance that turns the flame carmine-red is LiCl, the one that turns the flame lilac is KCl and the one that turns the flame apple green is $BaCl_2$.

2 The heat of the flame vaporises sodium chloride, producing some Na and Cl *atoms*. An electron is promoted into an orbital of the fourth shell in some of the sodium atoms. These electrons then fall back to the ground state, which is the 3*s*-orbital, and energy is given out in the form of light. The flame is yellow because the energy difference between the fourth shell orbital and the 3*s*-orbital is equivalent to the yellow line in the spectrum.

3 The magnesium atom is smaller than the calcium atom because magnesium has three shells of electrons and calcium has four shells. Thus, the outer *s* electron of magnesium is held more firmly and more energy is required to remove it.

4 a $CaO + H_2O \rightarrow Ca(OH)_2$
b $Mg(OH)_2 + 2HCl \rightarrow MgCl_2 + 2H_2O$

5 a $2Ca + O_2 \rightarrow 2CaO$ (or $Ca + ½O_2 \rightarrow CaO$)
b $Ca + 2H_2O \rightarrow Ca(OH)_2 + H_2$
c $Mg + Cl_2 \rightarrow MgCl_2$
d $Mg(s) + H_2O(g) \rightarrow MgO(s) + H_2(g)$

6 The white precipitate is insoluble magnesium hydroxide. Barium hydroxide is soluble and so is not precipitated.

7 Beryllium, as $BeCO_3$ is the least thermally stable of the group 2 carbonates. It decomposes readily on heating because the Be^{2+} ion is very polarising. The other group 2 ions are less polarising because they are larger.

8 a $4LiNO_3 \rightarrow 2Li_2O + 4NO_2 + O_2$
$2NaNO_3 \rightarrow 2NaNO_2 + O_2$
$2Mg(NO_3)_2 \rightarrow 2MgO + 4NO_2 + O_2$
b $Na_2CO_3 \rightarrow$ no reaction
$MgCO_3 \rightarrow MgO + CO_2$
$BaCO_3 \rightarrow$ no reaction

9 a amount of Na_2CO_3 weighed out = 1.33/106.0 = 0.01255 mol
b concentration of Na_2CO_3 solution = 0.01255/0.250 = 0.05019 mol dm^{-3}
c amount of Na_2CO_3 in titre = 0.05019 × 23.75/1000 = 0.001192 mol
d amount of HCl = 2 × 0.001192 = 0.002384 mol
e concentration of HCl solution = 0.002384/0.0250 = 0.0954 mol dm^{-3}

Hint

Either keep all figures on your calculator or cut to 4 s.f. (*not* 4 decimal places) and give your final answer to 3 s.f.

10 a amount of NaOH in titre = 0.200 × 26.80/1000 = 0.005360 mol
b amount of H_2X in 25.0 cm^3 = 0.5 × 0.005360 = 0.002680 mol
c amount of H_2X in 250 cm^3 = 10 × 0.002680 = 0.02680 mol
d molar mass of H_2X = mass/moles = 2.41/0.02680 = 89.9 g mol^{-1}

The periodic table: group 7 (chlorine to iodine)

1 First make a solution of the sample either in water or in dilute nitric acid. (If water is chosen, dilute nitric acid must then be added to destroy the carbonate). To this slightly acidic solution, add silver nitrate solution. If chloride is present, a white precipitate is produced that will dissolve in dilute aqueous ammonia.

2 H_2SO_4 is a *stronger* acid than HCl (or HI), and so it protonates the Cl^- ions in sodium chloride, producing HCl gas. HCl is *not* a strong enough reducing agent to reduce sulfuric acid, so no further reaction takes place. With sodium iodide, HI is produced by protonation of the iodide ions. However, HI is a strong reducing agent and it reduces the sulfuric acid and is itself oxidised to iodine.

3 Disproportionation is a reaction in which an element is simultaneously oxidised and reduced. Chlorine disproportionates when added to alkali:

$$\underset{(0)}{Cl_2} + 2OH^- \rightarrow \underset{(-1)}{Cl^-} + \underset{(+1)}{ClO^-} + H_2O$$

Chlorate(I) disproportionates when heated:

$$\underset{(+1)}{3ClO^-} \rightarrow \underset{(-1)}{2Cl^-} + \underset{(+5)}{ClO_3^-}$$

4 a amount (in moles) of thiosulfate in titre
= 0.105 × 22.75/1000 = 0.002389 mol

b amount (in moles) of liberated iodine
= ½ × 0.002389 = 0.001194 mol

c amount (in moles) of IO_3^- ions in 25 cm^3
= ⅓ × 0.001194 =0.0003981 mol

d amount (in moles) of potassium iodate in 250 cm^3
= 10 × 0.0003981 = 0.003981 mol
mass of *pure* potassium iodate = moles × molar mass
= 0.003981 × 214.0 = 0.8520 g

e purity of sample = 0.8520 × 100/1.00 = 85.2%

Kinetics

1 a The activation energy is the minimum energy that colliding molecules must have in order for them to react. It is measured in kJ mol^{-1}.

b A catalyst is a substance that speeds up a chemical reaction without being used up. It works by providing an *alternative* route with a lower activation energy.

Hint

Never say that a catalyst lowers the activation energy. Light is a form of energy. It is not a catalyst.

2 **Pressure:** as the pressure is increased, the number of molecules in a given volume increases. Therefore, the frequency of collision increases. The rate of reaction is dependent upon this and so it also increases.
Temperature: if the temperature is increased, the average kinetic energy of the molecules is also increased. This means that a greater proportion of the colliding molecules will have energy greater than or equal to the activation energy. So there will be a larger number of *successful* collisions and a faster rate of reaction.
Catalyst: this increases the rate by providing an alternative path with a lower activation energy, so a greater proportion of the colliding molecules have energy greater than or equal to the lower activation energy of the catalysed path.

3

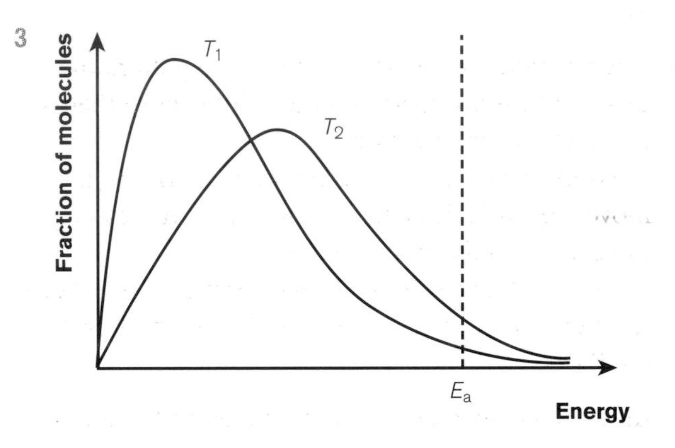

4 Bacterial and enzyme reactions are slowed down by a decrease in temperature for the same reason as ordinary chemical reactions are — fewer colliding molecules have the necessary activation energy to react on collision. So the low temperature in the refrigerator slows down the biological decay processes.

5 a

b

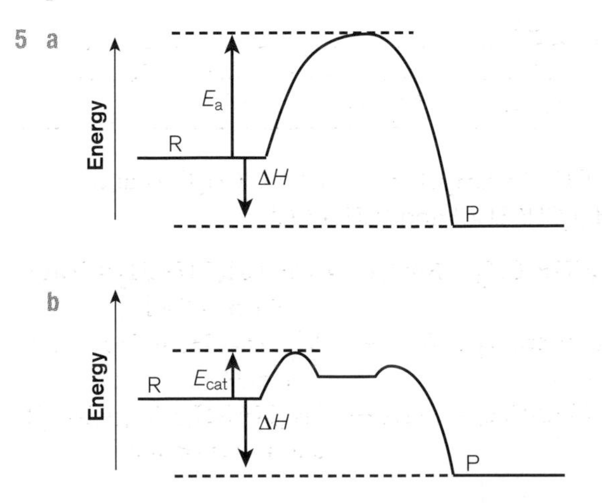

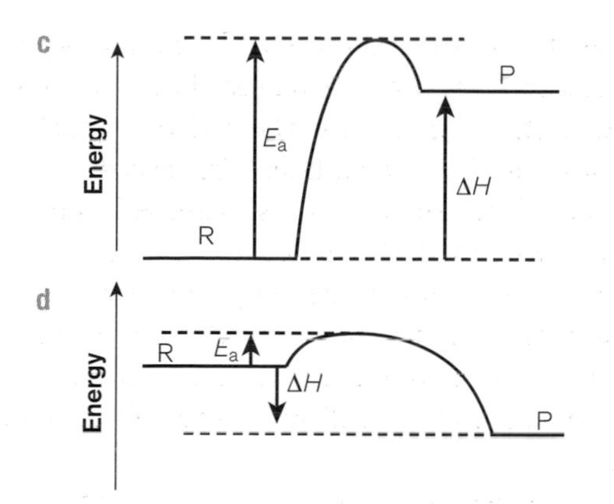

Chemical equilibria

1 a True
  b False
  c True

2 a A temperature decrease causes the equilibrium position to move from right to left because that is the exothermic direction.
  b Pressure increases, so the equilibrium position moves towards the left because there are fewer gas molecules on the left-hand side.
  c None. A catalyst speeds up the forward and reverse reactions equally, reducing the time taken to reach equilibrium.

3 a Alkali removes ethanoic acid, therefore the equilibrium position moves from right to left.
  b No change because ΔH is zero.

4 Solid dissolves completely because the high Cl^- concentration drives the equilibrium to the right.

Organic chemistry

1 a $CH_3CH_2CH_2OH$ and $CH_3CH(OH)CH_3$

> **Hint**
> This type of unambiguous formula is allowed unless the question asks for a *full* structural formula.

  b $CH_3CH_2CH_2CH_2Cl$ and $CH_3CHClCH_2CH_3$ and $(CH_3)_2CHCH_2Cl$ and $(CH_3)_3CCl$

2 a $CH_3CHBrCH_3 + NaOH \rightarrow CH_3CH(OH)CH_3 + NaBr$
  Propan-2-ol
  b $CH_3CHBrCH_3 + KOH \rightarrow CH_3CH{=}CH_2 + KBr + H_2O$
  Propene
  c $CH_3CHBrCH_3 + 2NH_3 \rightarrow CH_3CH(NH_2)CH_3 + NH_4Br$
  2-aminopropane

3 a CH_3CH_2COOH
  Propanoic acid
  Solution goes from orange to green
  b No reaction
  Solution stays orange
  c $CH_3CHClCH_3$
  2-chloropropane
  Steamy fumes

4 The C–Br bond is stronger than the C–I bond. This makes the activation energy higher, so the reaction is slower.

5 $C_6H_{10} + Br_2 \rightarrow C_6H_{10}Br_2$
5.67 g of C_6H_{10} = 5.67/82.0 = 0.0691 mol
As the reaction is a 1:1 reaction, the amount of product is also 0.0691 mol.
theoretical yield of $C_6H_{10}Br_2$ = 0.0691 × 242 = 16.7 g

$$\% \text{ yield} = \frac{15.4\text{ g}}{16.7\text{ g}} \times 100 = 92\%$$

Mechanisms

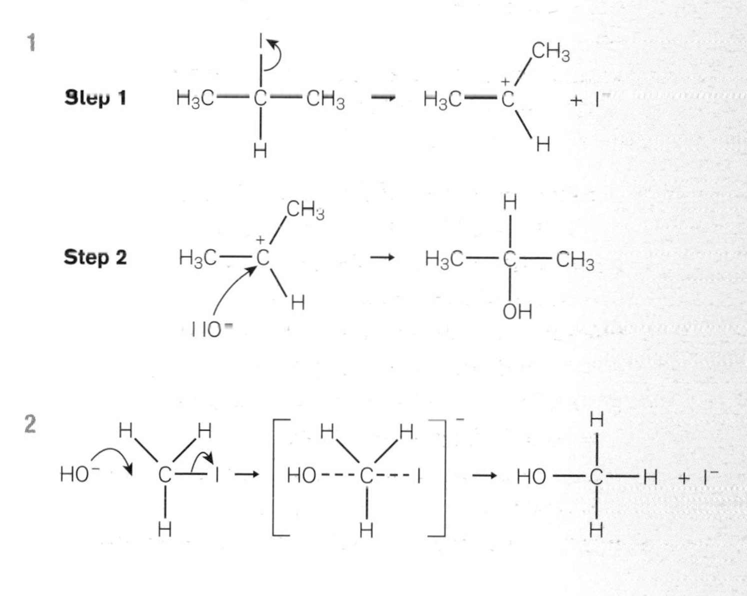

3 $Br{-}Br \xrightarrow{UV} 2Br\bullet$

Propagation: $Br\bullet + CH_3CH_3 \rightarrow HBr + CH_3CH_2\bullet$
$CH_3CH_2\bullet + Br_2 \rightarrow CH_3CH_2Br + Br\bullet$
Termination: $Br\bullet + Br\bullet \rightarrow Br_2$
or $CH_3CH_2\bullet + CH_3CH_2\bullet \rightarrow CH_3CH_2CH_2CH_3$
or $CH_3CH_2\bullet + Br\bullet \rightarrow CH_3CH_2Br$

4 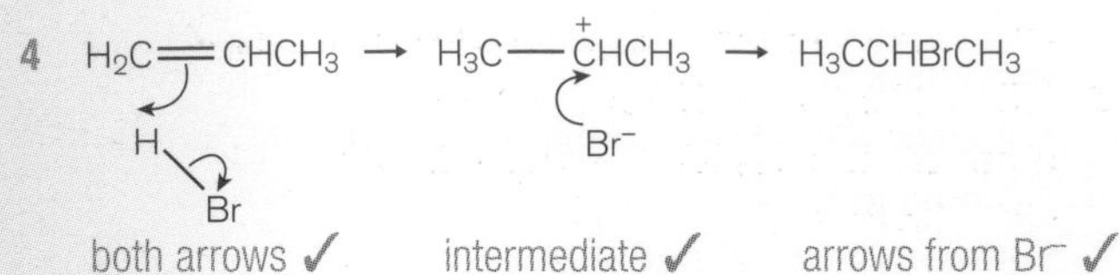

both arrows ✓ intermediate ✓ arrows from Br⁻ ✓

The secondary carbocation, $CH_3C^+HCH_3$ is more stable than the primary carbocation $^+CH_2CH_2CH_3$ that would lead to the minor product 1-bromopropane.

Mass spectra and IR

1 a *m/e* 124 $[C_3H_7{}^{81}Br]^+$; *m/e* 122 $[C_3H_7{}^{79}Br]^+$; *m/e* 29 $[C_2H_5]^+$
b There is no bromine isotope of mass 80.
c As it has a peak at *m/e* = 29, there is a CH_3CH_2 group in the molecule. Its structural formula is $CH_3CH_2CH_2Br$.
d As the relative atomic mass of bromine is 79.9, there must be about 50% of each isotope in natural bromine. Thus the peak heights are in the ratio of 1:1.

2 a NO is polar because nitrogen and oxygen have different electronegativities, so a stretching vibration will cause a change in dipole moment and so be infrared active.
b Cl_2 is non-polar, so the there is no change in dipole moment in a stretching vibration and so chlorine molecules do not absorb infrared radiation.
c Because chlorine is more electronegative than carbon, the bonds in CCl_4 are polar (although the molecule is symmetric and so is non-polar). A stretching vibration will cause a change in dipole moment, so it will absorb in the infrared.

3 a It is the IR spectrum of propan-1-ol, since there is no absorption at around $1700\,cm^{-1}$ due to the stretching of the C=O bond.
b The band at $3333\,cm^{-1}$ is caused by the hydrogen bonded O–H stretching; that at $2963\,cm^{-1}$ is due to the stretching of the C–H bond.

Hint

The frequency values in data tables are given as ranges because they represent the absorption of that bond in different molecules.

4 The band at $800–600\,cm^{-1}$ due to C–Cl will decrease and that at $3500–3300\,cm^{-1}$ due to N–H will increase (as will the band at $1040–1000\,cm^{-1}$ due to C–N).

5 A greenhouse gas is a substance that can absorb infrared radiation and which occurs in the gaseous state in the atmosphere.

Green chemistry

1 a atom economy

$$= \frac{\text{molar mass of product}}{\text{sum of molar masses of reactants}} \times 100$$

$$= \frac{64.5 \times 100}{46.0 + 208.5} = 25.3\%$$

b The reactants are ethanol and hydrogen chloride, which have a combined molar mass of 82.5. This is much less than that of ethanol and phosphorus pentachloride, so the atom economy is higher and the process is, therefore, preferable.

2 CH_4: yes. The bonds are polar and an asymmetric stretch will cause a change in dipole moment.
CI_4: no. The bonds are non-polar and so vibration will not result in a change in dipole moment.
H_2: no. The bond is non-polar.
CO: yes. The bond is polar and stretching will cause an increase in dipole moment.

3 Growing crops such as sugar, corn or wheat requires the use of fertilisers and fuel for ploughing and harvesting. Although fermentation does not require energy, distillation does, as does transporting the biofuel to filling stations. Therefore, significant amounts of carbon dioxide are emitted during the overall process.

4 In the stratosphere UV light breaks $CF_2ClCFCl_2$ homolytically:

$CF_2ClCFCl_2 + UV \rightarrow CF_2ClCFCl\bullet + Cl\bullet$

The Cl• radicals then react with ozone in a chain reaction:

$Cl\bullet + O_3 \rightarrow ClO\bullet + O_2$

ClO• radicals then react with the O• radical from the UV decomposition of ozone. Cl radicals are formed and so the chain reaction continues.

$ClO\bullet + O\bullet \rightarrow Cl\bullet + O_2$

Practice Unit Test 2

Section A

1 C	**5** B	**9** A	**13** D	**17** A
2 B	**6** A	**10** A	**14** B	**18** C
3 D	**7** D	**11** C	**15** B	**19** A
4 C	**8** B	**12** D	**16** B	**20** B

Section A total: 20 marks

Section B

21 a The nitrogen atom in NH_3 is smaller and more δ– than the phosphorus atom in PH_3 ✓, so it can form a hydrogen bond with another ammonia molecule, whereas phosphine cannot ✓. In phosphine, the strongest intermolecular forces are dispersion forces and these are weaker than the hydrogen bonds between ammonia molecules and so more energy is required to separate ammonia molecules, which means that ammonia has a higher boiling point ✓.

b As a HCl molecule has fewer electrons than a HBr molecule ✓ (18 rather than 35) it has weaker dispersion forces ✓ and a lower boiling point.

> **Hint**
> London, van der Waals or instantaneous induced dipole forces can be used in place of the term 'dispersion forces'.

c Both chloromethane and propane have the same number of electrons and so have similar strength London forces ✓. However, chloromethane is polar and so also has dipole–dipole forces ✓ and has, therefore, a higher boiling point.

d A Mg^{2+} ion has the same charge but a smaller radius than a Ca^{2+} ion ✓. Thus it polarises the CO_3^{2-} ion more ✓, so magnesium carbonate is easier to decompose.

e The C–I bond is weaker than the C–Cl bond ✓, so the activation energy is less ✓ and the rate is faster.

Total: 11 marks

22 a The reaction is exothermic ✓, so a higher temperature will cause the equilibrium position to shift to the left lowering the yield of sulfur trioxide ✓. A lower temperature will result in a rate that is too slow ✓, so a compromise temperature of 425°C is used.

b (i) amount of sodium hydroxide in titre
= concentration × volume in dm^3
= 0.111 × 27.45/100 = 0.003047 mol ✓

(ii) amount of sulfuric acid in 25 cm^3
= ½ × 0.003047 = 0.001523 mol ✓
amount in original sample = 10 × 0.001523
= 0.01523 mol ✓

(iii) mass of pure sulfuric acid in sample
= moles × molar mass
= 0.01523 × 98.1 = 1.494 = 1.49 g ✓

(iv) purity = 1.494 × 100/1.56 = 95.8% ✓

c (i) steamy fumes (of HCl vapour) ✓

(ii) steamy fumes ✓ and a red-brown gas ✓

d $KBr + H_2SO_4 \rightarrow KHSO_4 + HBr$ ✓
then $2HBr + H_2SO_4 \rightarrow Br_2 + 2H_2O + SO_2$
or $2KBr + 3H_2SO_4 \rightarrow$
$Br_2 + SO_2 + 2H_2O + 2KHSO_4$ ✓

e The reactant for the preparation of a bromoalkane is HBr and the yield is less because some is oxidised by the concentrated sulfuric acid ✓

Total: 14 marks

23 (a) (i) Bubbles of gas ✓ and the formation of a white solid ✓.

> **Hint**
> Do not say 'hydrogen evolved' as your observation. The nature of the gas is a deduction, *not* an observation.

(ii) $2C_2H_5OH + 2Na \rightarrow 2C_2H_5ONa + H_2$ ✓.

> **Hint**
> The halved equation giving $½H_2$ also would score the mark.

b (i) The orange solution goes green ✓.
(ii) The solution stays orange ✓.

c Propanoic acid ✓

d

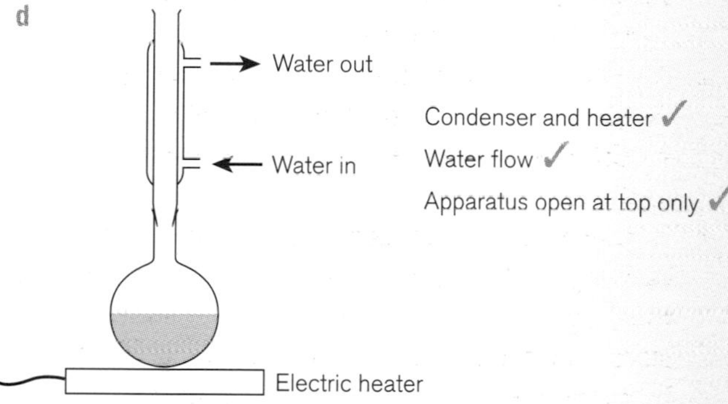

Total: 9 marks

24 a (i) $C_2H_5Br + KOH$ (or OH^-) $\rightarrow C_2H_5OH + KBr$ (or Br^-) ✓.

(ii) $C_2H_5Br + KOH \rightarrow CH_2{=}CH_2 + KBr + H_2O$ ✓.

b

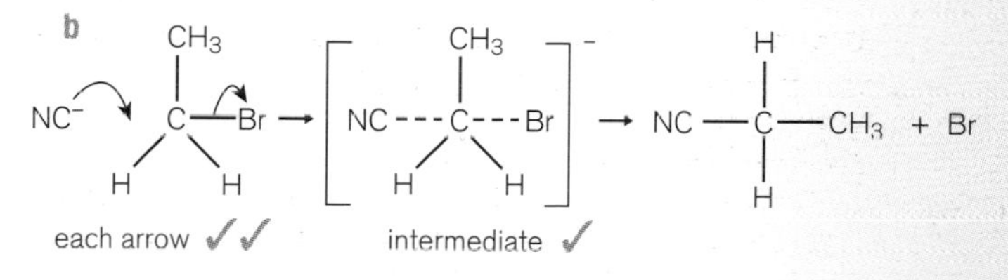

or

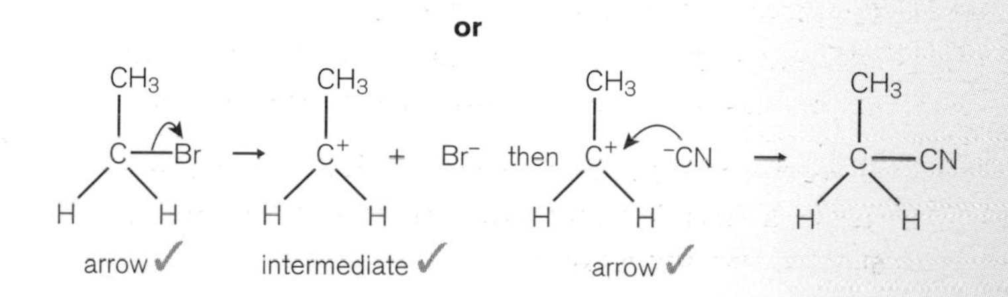

c Warm with aqueous sodium hydroxide ✓.
Add excess nitric acid ✓, then aqueous silver nitrate ✓.
A cream precipitate ✓, insoluble in dilute ammonia but soluble in concentrated ammonia ✓, confirms the presence of bromine in the organic compound.

Total: 10 marks

Section B total: 44 marks

Section C

25 (a) (i)  H ×• O •× H ✓ (with ×× above and below O) **(ii)**  ×× O ×× •• C •• ×× O ×× ✓

(b) (i) There are two lone pairs of electrons on the oxygen in H_2O ✓, and these repel the H–O bonding electrons ✓ forming a V-shaped molecule. There are no lone pairs on the carbon in CO_2, so only the C=O bonding electrons repel ✓ making the molecule linear.

(ii) Both molecules contain polar bonds ✓, but in O=C=O the dipoles cancel out because they are equal and opposite ✓. In water, the dipoles do not cancel as the molecule is V-shaped ✓.

c The (two) asymmetric ✓ stretching and bending vibrations cause a change in the dipole moment ✓ of the molecule and so it absorbs infrared radiation.

d A methane molecule absorbs infrared radiation to a greater extent than a carbon dioxide molecule ✓, but there is more carbon dioxide in the atmosphere ✓, so it has a greater impact on global warming ✓.

e CFCs are compounds that contain carbon, fluorine and chlorine and do not contain hydrogen ✓. In the stratosphere, they are decomposed by UV radiation to form chlorine radicals ✓. These destroy ozone in the ozone layer by a chain reaction ✓.

Any two of the following equations for the final 2 marks:

CF_2Cl_2 (or any other CFC) $\rightarrow$ $Cl\bullet + CF_2Cl\bullet$ (or equivalent)

$Cl\bullet + O_3 \rightarrow ClO\bullet + O_2$

$ClO\bullet + O\bullet \rightarrow Cl\bullet + O_2$ etc. in a chain process

Section C total: 16 marks

Paper total: 80 marks

About 56 marks would be needed for an A grade and about 27 for an E grade.

Unit 4 Rates, equilibria and further organic chemistry

Rates: how fast?

1 rate = $k[HI]^2$

$$k = \frac{\text{rate}}{[HI]^2}$$

$$= \frac{2.0 \times 10^{-4}\,\text{mol dm}^{-3}\,\text{s}^{-1}}{(0.050\,\text{mol dm}^{-3})^2} = 0.080\,\text{s}^{-1}\,\text{mol}^{-1}\,\text{dm}^3$$

2 Carry out the following procedure:

Place equal volumes of solution, e.g. 50 cm^3 of 0.10 mol dm^{-3} ethanoic acid and methanol, in flasks in a thermostatically controlled tank at 60°C.

Mix, start the clock and replace in the tank.

At intervals of time, pipette out 10 cm^3 portions and add to 25 cm^3 of iced water in a conical flask.

Rapidly titrate with standard sodium hydroxide solution using phenolphthalein as the indicator.

Repeat several times.

Plot a graph of the titre (which is proportional to the amount of ethanoic acid left) against time.

3 Time taken for the concentration to halve from 1.6 to 0.8 mol dm^{-3} = 26 minutes

Time taken to halve again to 0.4 mol dm^{-3} = 26 minutes

Time taken to halve again to 0.2 mol dm^{-3} = 26 minutes

$t_{½}$ is a constant, therefore the reaction is first order

4 Gradient $= \frac{-10 - (-6.0)}{0.0033 - 0.0028}$

$$= \frac{-4.0}{0.0005} = -8000 = -E_a/R$$

E_a = −gradient × R = −(−8000) × 8.3 = +66 000 J mol^{-1} = +66 kJ mol^{-1}

Entropy: how far?

1 a ΔS_{system} = 178 + (2 × 187) − (186 + (2 × 165)) = +36 J K^{-1} mol^{-1}

b ΔH = −124 + (2 × (−92)) −(−75) = −233 kJ mol^{-1}

$\Delta S_{surroundings}$ = $-\Delta H/T$ = −(−233 000/298) = +782 J K^{-1} mol^{-1}

c ΔS_{total} = ΔS_{system} + $\Delta S_{surroundings}$ = 36 + 782 = +818 J K^{-1} mol^{-1}

The value is positive, so the reaction is thermodynamically feasible.

2 $\Delta S_{solution}$ = −lattice energy + sum of hydration energies = −(−2237) + (−1650) + (2 × (−364)) = −141 kJ mol^{-1}

3 ΔS_{total} = ΔS_{system} − $\Delta H/T$

ΔS_{total}(AgF) = −21 + 20 000/298 = +46 J K^{-1} mol^{-1}

This is positive, so silver fluoride will be soluble.

ΔS_{total}(AgCl) = +34 −66 000/298 = −187 J K^{-1} mol^{-1}

This is negative, so silver chloride will be insoluble.

4 a Ca^{2+} has the same charge but a smaller radius than Ba^{2+}, so it has a stronger attraction for SO_4^{2-} ions and therefore a more exothermic lattice energy.

b Mg^{2+} has a greater charge and a smaller radius than Na^+, so it has a stronger attraction for the δ− oxygen in water and therefore it has a more exothermic hydration energy.

Equilibria and application of rates and equilibrium

1 $2SO_3 \rightleftharpoons 2SO_2 + O_2$

	SO_3	SO_2	O_2
Start (moles)	0.0200	0	0
Change (moles)	− 0.0058	+ 0.0058	+ 0.0029
Equilibrium moles	0.0142	0.0058	0.0029
Equilibrium concentration /mol dm^{-3}	0.0142/1.52 = 0.009342	0.0058/1.52 = 0.003816	0.0029/1.52 = 0.001908

Hint

1 mol of SO_3 gives 1 mol of SO_2 and ½ mol of O_2.

$$K_c = \frac{[SO_2]^2[O_2]}{[SO_3]^2}$$

$$= \frac{(0.003816)^2 \times 0.001908}{(0.009342)^2} = 3.18 \times 10^{-4}\,\text{mol dm}^{-3}$$

2 $2NO(g) + O_2(g) \rightleftharpoons 2NO_2(g)$

	NO	O_2	NO_2	Total
Start (moles)	1.0	1.0	0	2.0
Change (moles)	− 0.60	− 0.30	+ 0.60	
Equilibrium moles	0.40	0.70	0.60	1.70
Equilibrium mole fraction	0.40/1.7 = 0.234	0.70/1.7 = 0.409	0.60/1.7 = 0.351	
Partial pressure /atm	0.234 × 4 = 0.96	0.409 × 4 = 1.64	0.351 × 4 = 1.40	

$$K_p = \frac{p(NO_2)^2}{p(NO)^2\,p(O_2)}$$

$$= \frac{1.40^2}{0.936^2 \times 1.64} = 1.4\,\text{atm}^{-1}$$

3 a $\frac{[SO_3]_{initial}^2}{[SO_2]_{initial}^2[O_2]_{initial}} = \frac{(0.2)^2}{(0.2)^2 \times 0.1}$

$= 10\,mol^{-1}\,dm^3$

This is *not* equal to K_c, so the system is *not* at equilibrium. The quotient < K_c, so the reaction moves left to right, until equilibrium is reached.

The concentration expression only equals *K* when the system is in equilibrium.

b (i) As the reaction left to right is *exo*thermic, a *decrease* in temperature causes K_c to increase, so the position of equilibrium moves to the right until the quotient becomes equal to the larger K_c. Therefore more SO_2 is converted.

(ii) K_c is unaltered. As the reaction right to left results in an increase in the number of moles of gas, a decrease in pressure causes the quotient to become larger. Therefore, the position of equilibrium shifts from right to left reducing the quotient until it equals the unchanged K_c. Therefore less SO_2 is converted.

(iii) There will be no effect on either. A catalyst speeds up the rate of both forward and back reactions equally, so equilibrium is reached sooner.

(iv) There will be no effect on the value of K_c. Removal of sulfur trioxide makes the quotient smaller, so more sulfur dioxide and oxygen must react to increase the quotient until it again equals the unchanged K_c.

4 a $\Delta S_{total} = \Delta S_{system} - \Delta H/T$ and $\Delta S_{total} = R \ln K$ or $K = e^{\Delta S/R}$

At 25°C: $\Delta S_{total} = +215 - 124\,000/298$

$= -201\,J\,K^{-1}\,mol^{-1}$

$K = e^{-201/8.31} = 3.13 \times 10^{-11}$

At 400°C: $\Delta S_{total} = +215 - 124\,000/673$

$= +30.8\,J\,K^{-1}\,mol^{-1}$

$K = e^{+30.8/8.31} = 41$

b At 25°C , reaction is not noticeable. At 400°C, the reaction favours the products but both products and reactants are present in observable quantities.

Acid–base equilibria

1 $C_2H_5COOH \rightleftharpoons H^+ + C_2H_5COO^-$

$pK_a = 4.87$

$K_a = 10^{-4.87} = 1.35 \times 10^{-5}\,mol\,dm^{-3}$

$[H^+] = [C_2H_5COO^-]$ (as there is no other significant source of either ion)

$K_a = \frac{[H^+][C_2H_5COO^-]}{[C_2H_5COOH]} = \frac{[H^+]^2}{[C_2H_5COOH]}$

$= 1.35 \times 10^{-5}\,mol\,dm^{-3}$

$[H^+] = \sqrt{K_a[C_2H_2COOH]} = \sqrt{1.35 \times 10^{-5} \times 0.22}$

$= 1.72 \times 10^{-3}\,mol\,dm^{-3}$

$pH = -\log[H^+] = 2.76$

2 a

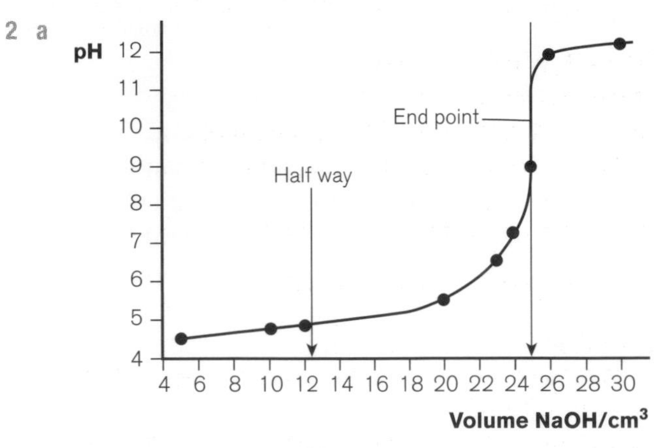

End point is at 25 cm^3 NaOH

Halfway to end point is 12.5 cm^3, which is when [HA] = [A⁻].

pH at this point = 4.95

$pK_a = 4.95$

b Phenolphthalein

Hint

This is a suitable indicator because its range of pH for *all* its colour change lies within the vertical part of the graph.

3 a It is a solution of known pH that resists a change in pH when small amounts of either acid or base are added. It consists of any weak acid and its conjugate base, for example ethanoic acid and sodium ethanoate.

Hint

Do *not* say that the pH is constant; nearly constant is acceptable.

b The salt is fully ionised and the weak acid is **slightly ionised**:

If H^+ ions are added to the solution, almost all of them are removed by reaction with the large reservoir of CH_3COO^- ions:

$H^+(aq) + CH_3COO^-(aq) \rightarrow CH_3COOH(aq)$

As $[CH_3COO^-]$ and $[CH_3COOH]$ are large relative to the small amount of H^+ added, their values remain almost the same — hence the pH hardly alters.

If OH^- ions are added, almost all of them are removed by reaction with the large reservoir of CH_3COOH molecules:

$OH^-(aq) + CH_3COOH \rightarrow CH_3COO^-(aq) + H_2O(l)$
As $[CH_3COO^-]$ and $[CH_3COOH]$ are large relative to the small amount of H^+ added, their values remain almost the same — hence the pH hardly alters.

Hint

The critical point is that both $[CH_3COOH]$ and $[CH_3COO^-]$ are large relative to the small amount of added H^+ or OH^- added. This is why a weak acid on its own is not a buffer.

4 $[H^+] = K_a$[weak acid]/[salt]
[weak acid] = 0.44 mol dm^{-3}
amount of salt = 4.4/82.0 = 0.0537 mol
∴ [salt] = 0.0537/0.1 = 0.537 mol dm^{-3}

$$[H^+] = \frac{1.74 \times 10^{-5} \times 0.44}{0.537} = 1.43 \times 10^{-5}\ \text{mol dm}^{-3}$$

∴ pH = −log (1.43 × 10^{-5}) = 4.85

Hint

The concentration and *not* the number of moles should be used in buffer calculations.
In both weak acid and buffer calculations, the same expression for K_a is used. For a weak acid, $[H^+] = [A^-]$, whereas in buffer calculations $[A^-]$ = [salt] and does *not* equal $[H^+]$.

Further organic chemistry

1

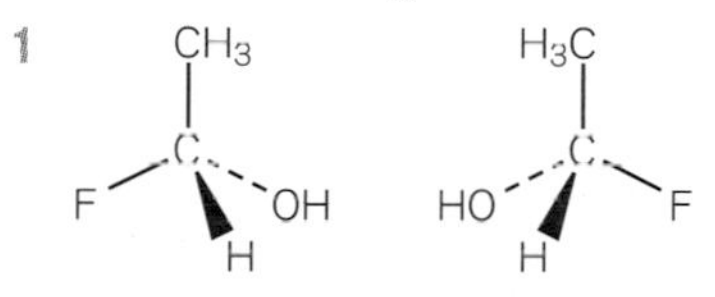

2 Step 1: $C_2H_5OH \rightarrow CH_3CHO$ — oxidise ethanol by warming with acidified potassium dichromate(VI) and distilling off the ethanal as it is formed
Step 2: $CH_3CHO \rightarrow CH_3CH(OH)CN$ — add hydrogen cyanide (HCN) in the presence of a catalyst of cyanide ions
Step 3: $CH_3CH(OH)CN \rightarrow CH_3CH(OH)COOH$ — hydrolyse by warming with dilute sulfuric acid

Hint

In answer to this type of question, you must identify reagents, special conditions and all intermediate compounds.

3

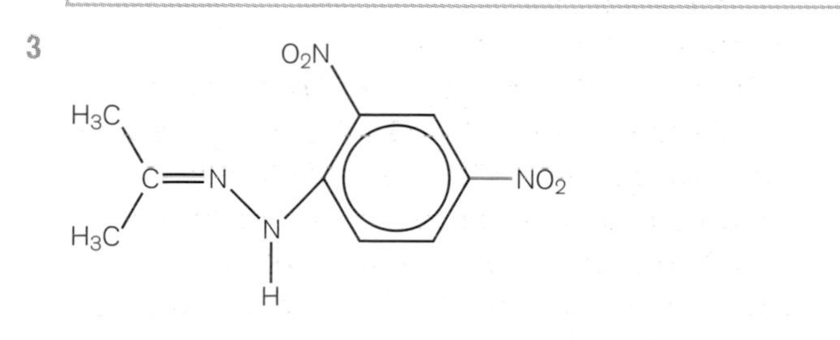

4

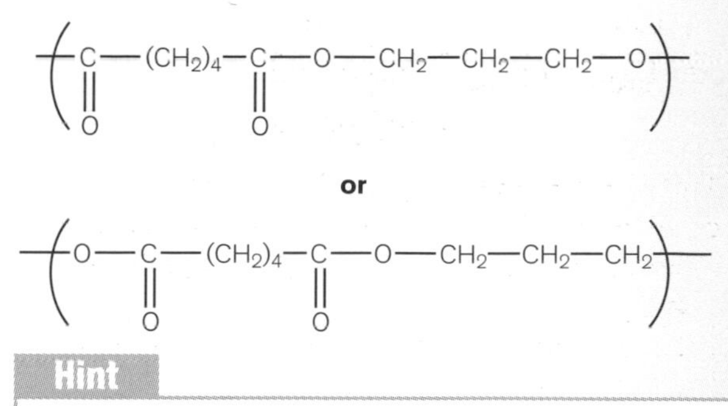

Hint

Check that there are four and *only* four oxygen atoms in the repeat unit of any polyester.

5 **a** Add water to each: ethanoyl chloride gives off steamy fumes of HCl; ethanoic acid gives no visible reaction.
Or add ethanol to each: ethanoyl chloride gives a product with a fruity ester-like smell; in the absence of a catalyst, ethanoic acid does not react.
Or add PCl_5 to each: ethanoic acid gives off steamy fumes of HCl; ethanoyl chloride gives no visible reaction.

b Add to each a solution of iodine and sodium hydroxide: propanone gives a pale yellow precipitate; propanal does not give a precipitate.
Add ammoniacal silver nitrate solution to each and warm: propanal gives a silver mirror; propanone does not give a mirror.

Spectroscopy and chromatography

1 Both have hydrogen nuclei in four different chemical environments. The splitting pattern for $CH_3CH_2CH_2OH$ is 3:6:3:1 and that for $CH_2=CHCH_2OH$ is 2:5:2:1. Therefore, the compound is 3-hydroxyprop-1-ene.

2 $C_4H_{10}O$ contains an OH group, so it must be an alcohol. It is not a tertiary alcohol because it can be oxidised by acidified dichromate. A secondary alcohol would give the ketone $CH_3COCH_2CH_3$ on oxidation. There are two possible primary alcohols, $CH_3CH_2CH_2CH_2OH$ and $(CH_3)_2CHCH_2OH$, that would give either an aldehyde or an acid on oxidation. The infrared data show that the oxidation product is not an acid, since there is no peak in the range 3200–3650 cm^{-1}. A ketone is not possible because the product gave a silver mirror with Tollens. The 1720 cm^{-1} band is probably due to the C=O in an aldehyde and the 2900 cm^{-1} peak due to alkyl C–H stretching.

The two most likely formulae for the oxidation product are $CH_3CH_2CH_2CHO$ and $(CH_3)_2CHCHO$. The former would have four lines in its NMR spectrum, so the oxidation product is $(CH_3)_2CHCHO$ and the original would be $(CH_3)_2CHCH_2OH$.

Practice Unit Test 4

Section A

1 D	**5** D	**9** B	**13** B	**17** C
2 D	**6** C	**10** B	**14** A	**18** B
3 A	**7** B	**11** C	**15** C	**19** C
4 C	**8** B	**12** D	**16** A	**20** B

Section A total: 20 marks

Section B

21 a (i) $\Delta S_{surroundings} = -\Delta H/T = -(-21\,600/298)$
$= +72.5\,J\,K^{-1}\,mol^{-1}$

converting to kelvin ✓ value and unit ✓

Hint

Remember to convert ΔH into $J\,mol^{-1}$ and °C to K.

(ii) $\Delta S_{system} = 259 + 187 - (201 + 161)$
$= +84\,J\,K^{-1}\,mol^{-1}$ ✓

(iii) $\Delta S_{total} = \Delta S_{system} + \Delta S_{surroundings}$
$= +156.5\,J\,K^{-1}\,mol^{-1}$ ✓
$= R\ln K$

$K = e^{\Delta S/R} = e^{156.5/8.31} = 1.51 \times 10^8$ ✓

(iv) Since ΔS_{total} is positive the reaction is thermodynamically feasible ✓. $K > 10^8$, so the reaction is effectively complete. ✓

b (i) $CH_3COCl + H_2O \rightarrow CH_3COOH + HCl$ ✓

(ii) $CH_3COCl + C_2H_5NH_2 \rightarrow C_2H_5NHCOCH_3 + HCl$ ✓

c (i) ΔS_{system} would be smaller because no gas is produced ✓.

(ii)

	CH_3COOH	C_2H_5OH	$CH_3COOC_2H_5$	H_2O
Initial moles	0.25	0.30	0	0.50
Change	−0.13	−0.13	+0.13	+0.13
Equilibrium moles	0.12	0.17	0.13	0.63 ✓
$[\,]_{eq}$	0.12/*V*	0.17/*V*	0.13/*V*	0.63/*V* ✓

$$K_c = \frac{[CH_3COOC_2H_5][H_2O]}{[CH_3COOH][C_2H_5OH]} ✓ = \frac{0.13/V \times 0.63/V}{0.12/V \times 0.17/V}$$

$= 4.0$ ✓

There is no need to know the volume as *V* cancels ✓.

(iii) *K* is unaltered as ΔH = zero ✓. As neither *K* nor the fraction (the quotient) is altered by a change in temperature, the position of equilibrium also remains unaltered ✓.

Total: 17 marks

22 a (i) 'Weak' means that the acid is only slightly ionised ✓.

Hint

Do not say a weak acid is partially ionised because that could mean over 50% ionised.

(ii) CN^- is the conjugate base ✓.

b (i) $K_a = \frac{[H^+][CN^-]}{[HCN]}$ ✓ $= \frac{[H^+]^2}{[HCN]}$

$[H^+] = \sqrt{K_a[HCN]} = \sqrt{4.0 \times 10^{-10} \times 1}$
$= 2.0 \times 10^{-5}$ ✓

$pH = -\log[H^+] = 4.70$ ✓

(ii) The first assumption is that $[H^+] = [CN^-]$. This is justifiable because there is no other source of CN^- ions and the H^+ from water is so small that it can be ignored ✓. The second assumption is that $[HCN]_{eq} = [HCN]_{initial} = 1.0\,mol\,dm^{-3}$. This is justifiable because HCN is a very weak acid so very little ionises (about 0.002%) ✓.

c $K_a = \frac{[H^+][salt]}{[acid]}$ ✓ and $[H^+] = 10^{-pH}$

$= 1 \times 10^{-9}\,mol\,dm^{-3}$ ✓

$$\frac{[CN^-]}{[HCN]} = \frac{[salt]}{[acid]} = \frac{K_a}{[H^+]} = \frac{4 \times 10^{-10}}{1 \times 10^{-9}} = 0.4{:}1 \text{ or } 1{:}2.5$$ ✓

d (i) Compare experiments 1 and 2: when $[CN^-]$ doubles and $[CH_3CHO]$ remains the same ✓, the rate doubles so the reaction is first order with respect to CN^- ions ✓.
Compare experiments 2 and 3: when $[CH_3CHO]$ increases by a factor of four and $[CN^-]$ remains constant, the rate increases by a factor of four, so the reaction is first order with respect to CH_3CHO ✓.

Hint

You must either mention the two experiments that you are comparing or say that when one concentration changes, the other remains constant.

(ii)

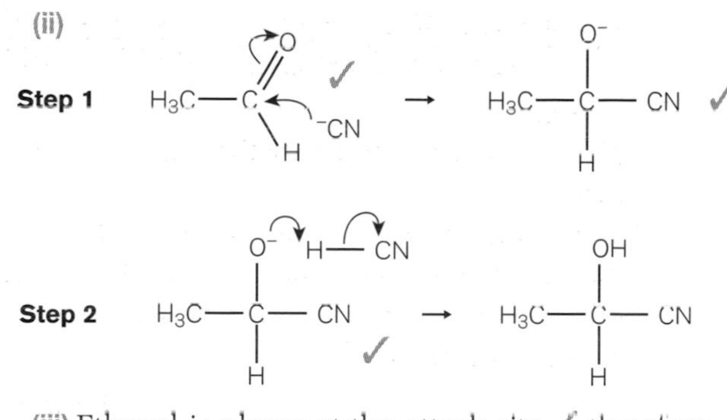

(iii) Ethanal is planar at the attack site ✓, therefore the CN^- ions can attack from either above or below the plane and the result is a racemic mixture ✓.

Total: 18 marks

23 a Add dilute sulfuric acid and potassium dichromate(VI) ✓ to the acrolein and warm under reflux ✓. Distil off the product ✓.

> **Hint**
> 'Describe' requires a fuller account than 'outline', which only needs the reagents and conditions to be identified.

b Propenoic acid has a δ+ hydrogen in the –COOH group which can hydrogen bond with the δ– oxygen in another propenoic acid molecule ✓. Acrolein does not have a sufficiently δ+ hydrogen so only forms the weaker van der Waals (and dipole–dipole) forces ✓.

> **Hint**
> The terms dispersion, London or induced dipole can be used in place of van der Waals.

c (i) $CH_2{=}CHCH_2OH$ ✓
(ii) CH_3CH_2COOH ✓

d (i) $CH_2{=}CHCOOH + PCl_5 \rightarrow CH_2{=}CHCOCl + POCl_3 + HCl$ ✓
(ii) $2CH_2{=}CHCOOH + Na_2CO_3 \rightarrow 2CH_2{=}CHCOONa + CO_2 + H_2O$ ✓

e The CHO hydrogen in acrolein will have a peak at δ = 9.1 ✓ (see data booklet). There is one hydrogen atom on a a neighbouring carbon atom and so will be split, according to the (*n* + 1) rule into two ✓. The COOH hydrogen in propenoic acid will have a peak at δ = 12 ✓. This will not be split because there is no hydrogen on the neighbouring carbon atom (the C of the COOH) ✓.

f

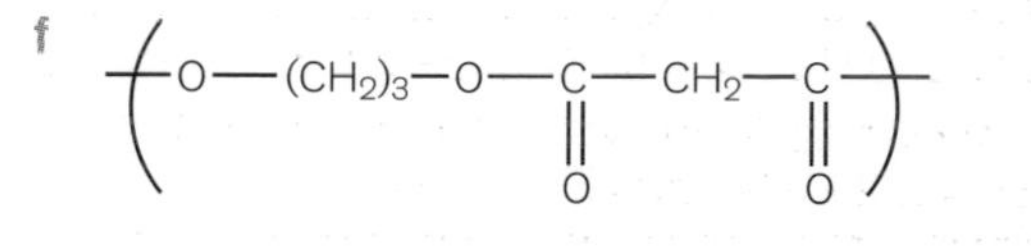

or

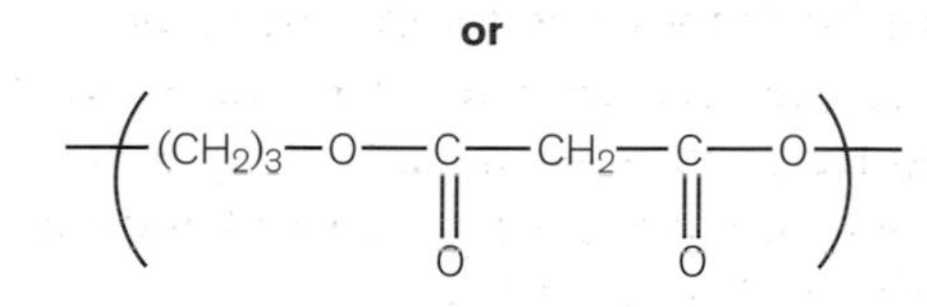

Marks for the diagram: ester link drawn out showing C=O ✓; rest of unit including 'continuation' bonds ✓.

> **Hint**
> The repeat unit of a polyester has four oxygen atoms.

Total: 15 marks

Section B total: 50 marks

Section C

24 a

Element	%	Divide by r.a.m.	Divide by smallest
C	66.7	66.7/12 = 5.56	5.56/1.39 = 4.0
H	11.1	11.1/1 = 11.1 ✓	11.1/1.39 = 8.0
O	22.2	22.2/16 = 1.39	1

Therefore empirical formula = C_4H_8O ✓

b Relative molecular mass = 72; relative empirical mass = 72 ✓. Therefore, molecular formula is C_4H_8O ✓.

c C=C (alkene) ✓; C=O (carbonyl or both of aldehyde and ketone) ✓; OH (alcohol) ✓.

d (i) It does not contain a C=C group ✓.
(ii) It does not contain an OH group (is not an alcohol) ✓.
(iii) It is an aldehyde or a ketone ✓.
(iv) There is no CH_3CO group ✓.

> **Hint**
> Do *not* say that it does not contain a $CH_3CH(OH)$ group as an alcohol has been ruled out in (ii).

(v) It is an aldehyde ✓.

e

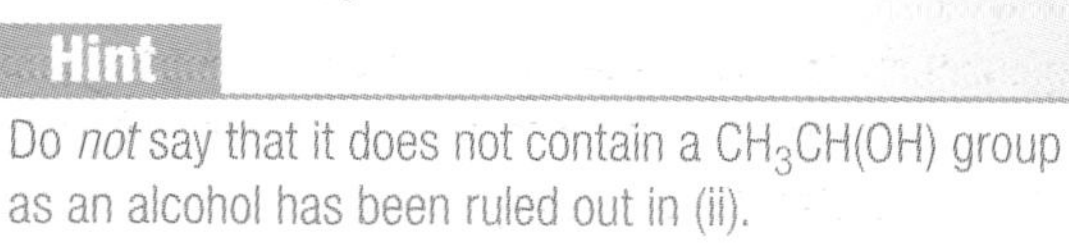

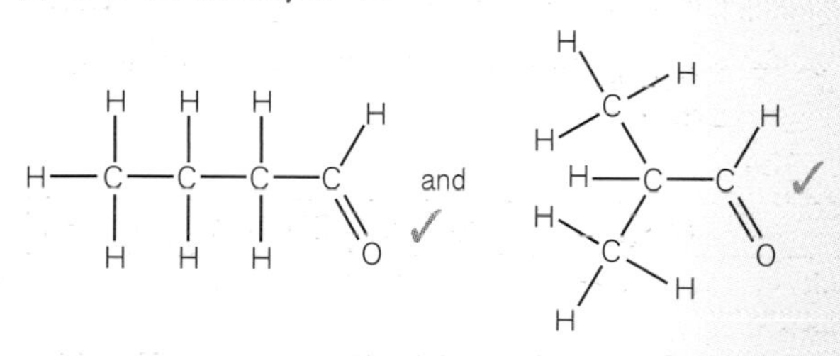

f (i) Either the $C_2H_5^+$ ion or the CHO^+ ion ✓.
(ii) $(CH_3)_2CHCHO$ has hydrogen nuclei in three different environments thus having three lines in its NMR spectrum ✓. $CH_3CH_2CH_2CHO$ has hydrogens in four environments and therefore has four lines in its NMR spectrum and is **X** ✓.

g $\Delta H_f = 4 \times (-394) + (4 \times (-286)) - (-2476)$
$= -244\ kJ\ mol^{-1}$ ✓
Use of × 4 (twice) ✓. Correct signs ✓.

Total: 20 marks

Paper total: 90 marks

About 65 marks would be needed for an A grade and about 33 for an E grade.

Unit 5 Transition metals and organic nitrogen chemistry

Redox equilibria

1 a (i) $Cr_2O_7^{2-}(aq) + 14H^+(aq) + 6e^- \rightarrow 2Cr^{3+}(aq) + 7H_2O$

(ii) $Sn^{4+}(aq) + 2e^- \rightarrow Sn^{2+}(aq)$

(iii) $IO_3^-(aq) + 6H^+(aq) + 5e^- \rightarrow ½I_2 + 3H_2O$

(iv) $I_2 + 2e^- \rightarrow 2I^-(aq)$

> **Hint**
> All are reductions and so have electrons on the left-hand side of the equation. The number of electrons equals the total change in oxidation number.

b (i) $Cr_2O_7^{2-}(aq) + 14H^+(aq) + 3Sn^{2+}(aq) \rightarrow 2Cr^{3+}(aq) + 7H_2O + 3Sn^{4+}(aq)$

$E_{cell} = +1.33 - 0.15 = +1.18\,V$

The reaction is feasible because $E_{cell} > 0$.

> **Hint**
> Equation a (ii) is reversed, multiplied by 3, and added to equation a (i).

(ii) $IO_3^-(aq) + 6H^+(aq) + 5I^-(aq) \rightarrow 3I_2 + 3H_2O$

$E_{cell} = +1.19 - 0.54 = +0.65\,V$

The reaction is feasible because $E_{cell} > 0$.

> **Hint**
> Equation a (iv) is reversed, multiplied by 5/2, and added to equation a (iii).

2 $Fe(s) + 2H^+(aq) \rightarrow Fe^{2+}(aq) + H_2(g)$

$5Fe^{2+}(aq) + MnO_4^-(aq) + 8H^+(aq) \rightarrow 5Fe^{3+}(aq) + Mn^{2+}(aq) + 4H_2O$

amount of $MnO_4^- = 0.0235 \times 0.0200 = 4.70 \times 10^{-4}$ mol

amount of Fe^{2+} in 25 cm³ sample $= 4.70 \times 10^{-4} \times 5/1 = 0.00235$ mol

> **Hint**
> 5/1 because there are $5Fe^{2+}$ to $1MnO_4^-$ in the equation.

amount of Fe^{2+} in 250 cm³ sample = 0.0235 mol

mass of Fe^{2+} = mass of Fe = 0.0235 × 55.8 = 1.311 g

$$\text{purity of iron in steel} = \frac{1.311}{1.320} \times 100 = 99.3\%$$

3 a $2Cr^{3+}(aq) + Zn(s) \rightarrow 2Cr^{2+}(aq) + Zn^{2+}(aq)$

$E^{\ominus}_{cell} = -0.41 + 0.76 = +0.35\,V$

b $2Cr^{3+}(aq) + 3Zn(s) \rightarrow 2Cr(s) + 3Zn^{2+}(aq)$

$E^{\ominus}_{cell} = -0.90 + 0.76 = -0.14\,V$

> In both, the reactants are Zn and Cr^{3+}, so the Cr^{3+} half-equations are left unaltered and the Zn^{2+}/Zn half-equation is reversed and the sign of its $E^{\ominus}$ changed. Note that the stoichiometry has to be changed so that the numbers of electrons cancel.

c The $E^{\ominus}_{cell}$ for **b** is negative, so the reaction is not feasible. $E^{\ominus}_{cell}$ for **a** is positive, so the reaction is feasible. The products are Cr^{2+} and Zn^{2+}.

Transition metal chemistry

1 The water ligands split the *d*-orbitals into two of higher energy and three of lower energy. When white light is passed through, an electron absorbs light energy and moves from a lower to a higher energy level, removing some of the red/green light and leaving blue light.

2 a $[Cr(H_2O)_6]^{3+}(aq) + 3OH^-(aq) \rightarrow Cr(OH)_3(s) + 6H_2O(l)$

> **Hint**
> This reaction is deprotonation.

then $Cr(OH)_3(s) + 3OH^-(aq) \rightarrow Cr(OH)_6^{3-}(aq)$

> **Hint**
> This occurs because Cr(III) is amphoteric.

b $[Fe(H_2O)_6]^{2+}(aq) + 2OH^-(aq) \rightarrow Fe(OH)_2(s) + 6H_2O(l)$

then no further reaction

c $[Zn(H_2O)_4]^{2+}(aq) + 2OH^-(aq) \rightarrow Zn(OH)_2(s) + 4H_2O(l)$

then $Zn(OH)_2(s) + 2OH^-(aq) \rightarrow [Zn(OH)_4]^{2-}(aq)$

3 a $[Fe(H_2O)_6]^{3+}(aq) + 3NH_3(aq) \rightarrow Fe(OH)_3(s) + 3NH_4^+(aq) + 3H_2O(l)$

> **Hint**
> This reaction is deprotonation.

then no further reaction

b $[Cu(H_2O)_6]^{2+}(aq) + 2NH_3(aq) \rightarrow Cu(OH)_2(s) + 2NH_4^+(aq) + 4H_2O(l)$

> **Hint**
> This reaction is also deprotonation.

then $Cu(OH)_2(s) + 4NH_3(aq) + 2H_2O(l) \rightarrow [Cu(NH_3)_4(H_2O)_2]^{2+}(aq) + 2OH^-(aq)$

> **Hint**
> The overall reaction is ligand exchange.

4 a The copper ions in hydrated copper(II) sulfate have four water ligands. These cause the *d*-orbitals to split. When the crystals are exposed to light, some of the frequencies are absorbed and an electron is promoted to one of the higher *d*-orbitals. There are no ligands in anhydrous copper(II) sulfate, so the *d*-orbitals are not split and promotion of an electron is impossible. The electron configuration of copper in the copper(I) complex is [Ar] $4s^0\ 3d^{10}$. Although the *d*-orbitals are split by the ligands, the orbitals are full, so promotion is impossible.

b If diaminomethane were a bidentate ligand, it would have to form a ring of four atoms, whereas 1,2-diaminoethane would form a ring of five atoms. A four-membered ring is too strained; a five-membered ring is not strained.

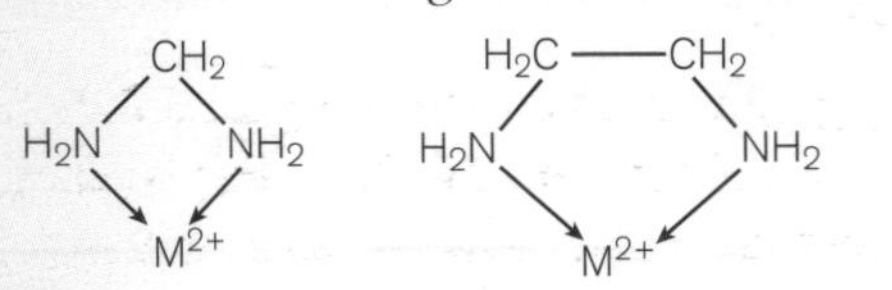

Organic chemistry: arenes

1 a Bromine reacts with the iron catalyst to form $FeBr_3$:

$$2Fe + 3Br_2 \rightarrow 2FeBr_3$$

The $FeBr_3$ then reacts with more bromine to form Br^+ (the electrophile) and $FeBr_4^-$.

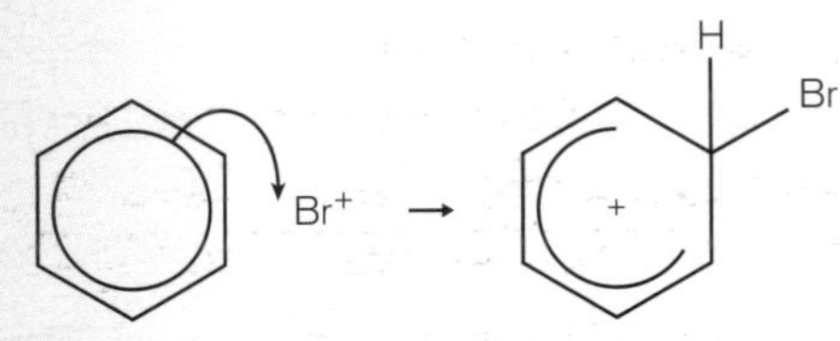

The intermediate cation loses an H^+ to $FeBr_4^-$.

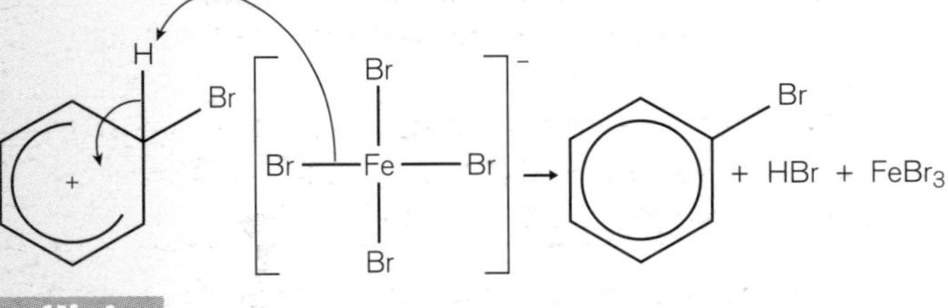

Hint

Check that:
- ★ the arrow starts on the delocalised ring and goes towards the Br of Br^+
- ★ the intermediate has a broken delocalised ring with a + inside it
- ★ the arrow starts from the σ-bond of the ring/H atom and goes inside the hexagon (but not directly to the +)

b It is energetically favourable for the intermediate cation to lose an H^+ and regain the stability of the benzene ring, rather than to add Br^- (as happens with alkenes).

2 The lone pair of p_z-electrons on the oxygen atom overlaps with the delocalised π-electrons of the six carbon atoms. This causes the electron density in the ring to increase and makes phenol more susceptible to electrophilic attack.

Organic nitrogen compounds

1 a $[Cu(NH_2C_6H_5)_4(H_2O)_2]^{2+}$

b $C_6H_5NHC_2H_5$

c $C_6H_5NHCOCH_3$

Hint

The chloroethane and the ethanoyl chloride react with the NH_2 group and not the benzene ring.

2 a $^+H_3NCH(CH_3)COOH$

b $H_2NCH(CH_3)COO^-$

c $H_2NCH(CH_3)COOC_2H_5$

3 a At less than 0°C the reaction is too slow. Above 10°C the product, benzene diazonium chloride, decomposes.

b

4 a

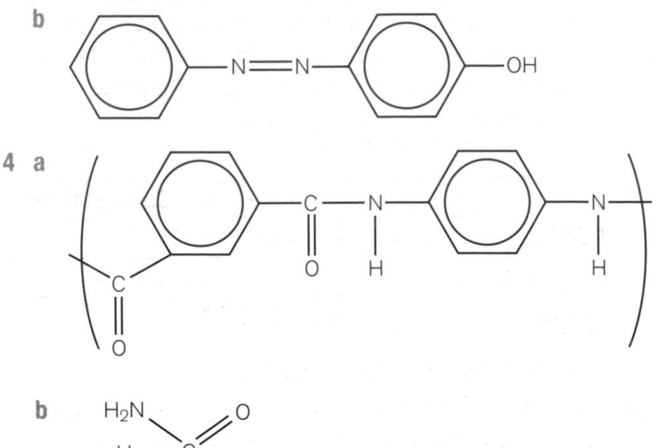

b

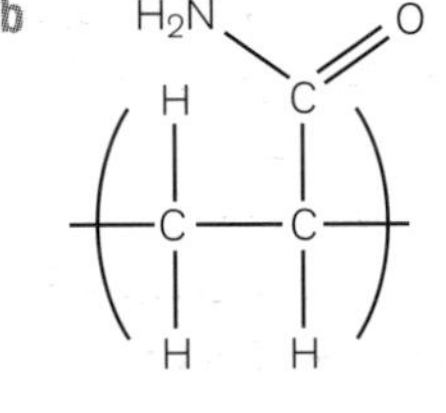

Organic analysis and synthesis

1 a Add Tollens' reagent and warm. Pentanal gives a silver mirror; the solution remains colourless with pentan-3-one.

or

Add Fehling's (or Benedict's) solution and warm. Pentanal gives a red precipitate; the solution remains blue with pentan-3-one.

b Heat under reflux with aqueous sodium hydroxide.

Hint

Halogenoalkanes are covalent and so must first be hydrolysed to produce halide ions.

Cool and acidify with dilute nitric acid, than add silver nitrate solution. The bromo-compound gives a cream precipitate that is insoluble in dilute ammonia but soluble in concentrated ammonia. The chloro-compound gives a white precipitate that dissolves in dilute ammonia.

c Warm with dilute sulfuric acid and potassium dichromate(VI) solution and distil off any product into Tollens' reagent. Methylpropan-2-ol does not change the colour of the potassium dichromate(VI). Methylpropan-1-ol turns it from orange to green and the distillate gives a silver mirror. Butan-2-ol turns it from orange to green and the distillate has no effect on Tollens' reagent. To confirm, add a few drops of the butan-2-ol to iodine and aqueous sodium hydroxide and warm gently. A yellow precipitate of iodoform will be produced.

Hint

The primary alcohol is partially oxidised to an aldehyde. The secondary alcohol is oxidised to a ketone.The tertiary alcohol is not oxidised. Butan-2-ol contains the $CH_3CH(OH)$ group and so gives a positive iodoform test.

2 a

	%	Divide by r.a.m.	Divide by smallest
Carbon	80.0	80.0/12.0 = 6.67	6.67/0.83 ≈ 8
Hydrogen	6.7	6.7/1.0 = 6.7	6.7/0.83 ≈8
Oxygen	13.3	13.3/16.0 = 0.83	1

The empirical formula is C_8H_8O.

b It does not have a C=C group: it has a C=O group (it is an aldehyde or a ketone); the high C to H ratio implies that it has a benzene ring.

c The peak at 120 is due to the molecular ion $C_8H_8O^+$; that at 105 is caused by the loss of CH_3 and that at 77 caused by the loss of CH_3CO. Therefore, the compound is $C_6H_5COCH_3$.

3 3200 cm^{-1} due to O–H; 1720 cm^{-1} due to C=O; 1150 cm^{-1} due to C–O.

4 Six of the carbon atoms are in a benzene ring. The remainder are either a C_2H_5 group or two CH_3 groups.
Y is $C_6H_5CH_2CH_3$ — five hydrogens in C_6H_5, 2 hydrogens in CH_2 and 3 hydrogens in CH_3.
Z is $C_6H_4(CH_3)_2$ — six hydrogens in two CH_3 groups, four hydrogens in C_6H_4.

5 Step 1: $C_2H_5OH \rightarrow CH_3CHO$ — warm with acidified potassium dichromate(VI) and distil off the product as it forms.
Step 2: $CH_3CHO \rightarrow CH_3CH(OH)CN$ — add HCN with a catalyst of KCN.
Step 3: $CH_3CH(OH)CN \rightarrow CH_3CH(OH)COOH$ — heat under reflux with dilute sulfuric acid.

Practice Unit Test 5

Section A

1 D	**5** A	**9** A	**13** D	**17** C
2 B	**6** A	**10** B	**14** B	**18** A
3 B	**7** C	**11** D	**15** A	**19** D
4 C	**8** D	**12** C	**16** A	**20** B

Section B

21 a (i) Dip a platinum electrode ✓ into a solution that is 1 mol dm^{-3} for MnO_4^-, Mn^{2+} **and** H^+ ✓. Connect via a salt bridge ✓ to a standard hydrogen electrode ✓.

(ii) It is not possible to measure a potential directly; a potential difference is measured ✓.

b (i) Multiply the MnO_4/Mn^{2+} half-equation by 2; reverse and multiply the $CO_2/(COOH)_2$ half-equation by 5 ✓ and then add them together.

$2MnO_4^- + 16H^+ + 10e^- \rightarrow 2Mn^{2+} + 8H_2O \quad E^\ominus = +1.52\,V$

$5(COOH)_2 \rightarrow 10CO_2 + 10H^+ + 10e^- \quad E^\ominus = (-0.49)\,V$

$2MnO_4^- + 6H^+ + 5(COOH)_2 \rightarrow 2Mn^{2+} + 10CO_2 + 8H_2O$ ✓

$E^\ominus_{cell} = 1.52 + 0.49 = +2.01\,V$ ✓

Hint

When a half-equation is reversed, the sign of $E^\ominus$ must be changed.
When a half-equation is multiplied, the $E^\ominus$ value is not altered.
On adding the half-equations, you must cancel *all* the electrons and as many of the H^+ as possible.

(ii) The activation energy is too high ✓.
(iii) +3 ✓

c amount (in moles) of $KMnO_4$
= concentration × volume in dm^3
= 0.0225 × 26.75/1000 = 6.019 × 10^{-4} mol ✓

amount (in moles) of $(COOH)_2$ = in 25 cm^3
= 5/2 × 6.019 × 10^{-4} = 0.001505 mol ✓

amount (in moles) in 1 dm^3 = 40 × 0.001505
= 0.06019 mol ✓

concentration in g dm^{-3} = moles × molar mass
= 0.06019 × 90
= 5.42 g dm^{-3} ✓

Total: 14 marks

22 a Cu^+: $1s^2\,2s^2\,2p^6\,2s^2\,2p^6\,3d^{10}$
Cu^{2+}: $1s^2\,2s^2\,2p^6\,2s^2\,2p^6\,3d^9$ ✓

b The extra energy required to remove one 3*d*-electron ✓ is more than compensated for by the extra energy released in forming the lattice or in hydrating the 2+ ion ✓.

c (i) It is a metal that has an unpaired *d*-electron in at least one of its ions ✓.

(ii) A ligand that is capable of forming two dative bonds with a metal ion ✓.

(iii) The number of dative bonds from ligands to the metal ion ✓.

d (i) The water ligands cause the *d*-orbitals to split into different energy levels ✓. When light is passed through, some of the frequencies are absorbed ✓ and an electron is promoted from a lower to a higher split *d*-orbital ✓.

(ii) The ligands are different so the energy gap of the split orbitals is different and so different frequencies are absorbed ✓.

(iii) There are no ligands ✓, so there is no splitting ✓.

(iv) The copper(I) ion has a *full* d^{10} configuration and so (although the *d*-orbitals are split) no promotion is possible ✓.

e $[Cu(H_2O)_6]^{2+} + 3NH_2CH_2CH_2NH_2 \rightarrow [Cu(NH_2CH_2CH_2NH_2)_3]^{2+} + 6H_2O$ ✓

f CuCl forms $[Cu(NH_3)_4]^+$ ✓ in complex formation. ✓ $[Cu(NH_3)_4]^+$ forms $[Cu(NH_3)_4(H_2O)_2]^{2+}$ ✓ in oxidation (by oxygen in air) ✓

Total: 18 marks

23 a (i) An electrophile is a species that attacks electron-rich (δ–) sites ✓ and forms a bond by accepting a pair of electrons ✓.

(ii) $C_2H_2Cl + AlCl_3 \rightarrow CH_3CH_2^+ + AlCl_4^-$ ✓

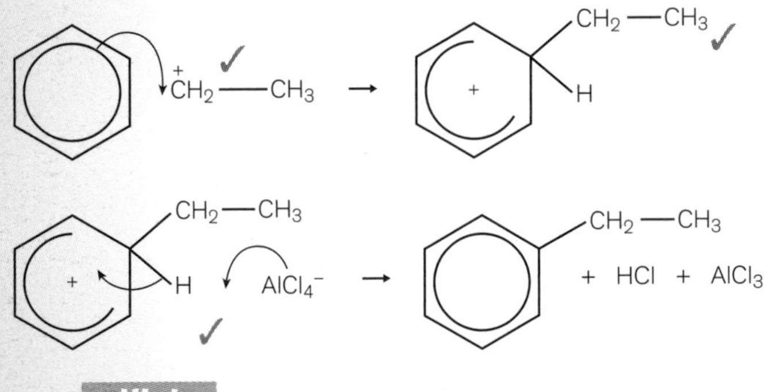

Hint

Make sure that the charge of the electrophile is on the CH_2 and not the CH_3 group, and that the bond from the ring goes to the CH_2 and not the CH_3.
Do not forget the charge on the intermediate.

b (i) $C_6H_6 + HNO_3 \rightarrow C_6H_5NO_2 + H_2O$ ✓

(ii) Below 50°C the reaction is too slow ✓and above 60°C a significant amount of dinitrobenzene is formed ✓.

(iii) The lower layer because its density is more than 1 g dm^{-3} ✓.

(iv) To remove acid impurities ✓.

(v) To dry the nitrobenzene ✓.

c (i) Tin (or iron) ✓ and concentrated hydrochloric acid ✓

(ii) $C_6H_5NH_2 + H^+ \rightarrow C_6H_5NH_3^+$ ✓
$C_6H_5NH_2 + CH_3COCl \rightarrow C_6H_5NHCOCH_3 + HCl$ ✓

(iii) $C_6H_5NH_2 + 2H^+ + NO_2^- \rightarrow C_6H_5N_2^+ + 2H_2O$ ✓

(iv)

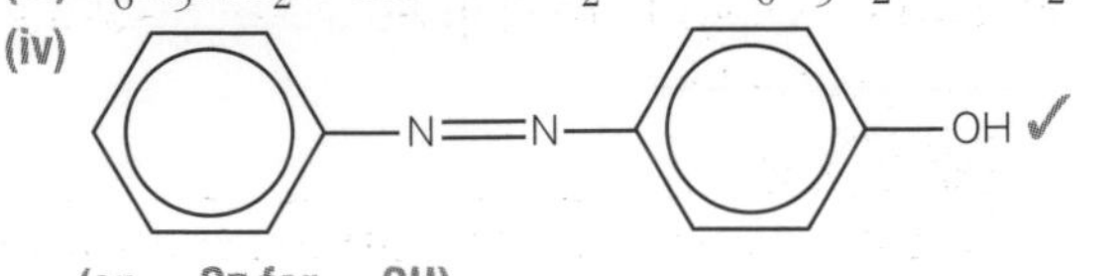

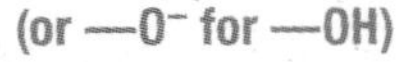
(or —O⁻ for —OH)

Total: 18 marks

Section C

24 (a) (i)

	%	Divide by r.a.m.	Divide by smallest
Carbon	47.06	47.06/12.0 = 3.92	3.92/2.94 = 1.33
Hydrogen	5.88	5.88/1.0 = 5.88	5.88/2.94 = 2
Oxygen	47.06	47.06/16.0 = 2.94	1

✓

Ratio is 1.33:2:1 or multiplying by 3 to get integers is 4:6:3 ✓

Empirical formula = $C_4H_6O_3$

(ii) PCl_5: **Z** contains an –OH group (is an alcohol or an acid or both) ✓
Na_2CO_3: **Z** contains a –COOH group ✓
Brady's: **Z** contains a >C=O group (is an aldehyde or a ketone) ✓
Tollens': **Z** contains a –CHO group ✓

(iii) amount of sodium hydroxide in titre = 0.105 × 22.05/1000 = 0.002315 mol ✓
amount of **Z** in 25.0 cm^3 = 0.002315 mol
amount of **Z** in 250 cm^3 = 10 × 0.002315 = 0.02315 mol ✓
molar mass of **Z** = mass/moles = 2.36/0.02315 = 102 g mol^{-1} ✓
mass of empirical formula = 102, so molecular formula is $C_4H_6O_3$ ✓

(iv) $CH_2(CHO)CH_2COOH$ ✓ or $CH_3CH(CHO)COOH$ ✓

(v) Both have four peaks in their NMR spectra, but the first compound will have a splitting pattern of 1, 3, 3 and 4, and so is **Z** ✓, whereas the second will have a splitting pattern of 1, 2, 2 and 5 and so is not **Z** ✓.

b (i)

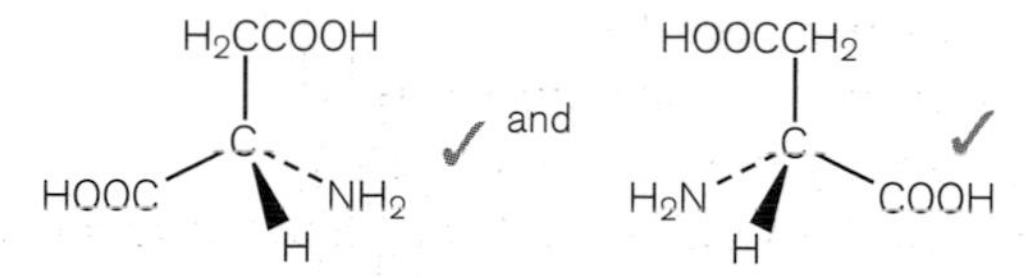

One will rotate the plane of plane polarised light clockwise and the other will rotate it anticlockwise ✓.

(ii) Two half-lives have elapsed ✓ so the skull is 500 years old ✓.

Hint

Assume that four units of the L-isomer were present initially. After one half-life there will be two units of L-aspartic acid and two of D-aspartic acid. After another half-life, there will be one of L-aspartic acid and three of D-aspartic acid.

(iii) The error is ±50 years per half-life, so the error in dating the skull is ±100 years ✓.

Total: 20 marks

Paper total: 90 marks

About 65 marks would be needed for an A grade and about 33 for an E grade.

Appendix

Organic reactions

Unit 1

Alkanes, e.g. ethane C_2H_6

Combustion to carbon dioxide and water

Reactant: oxygen (air)
Equation: $2C_2H_6 + 7O_2 \rightarrow 4CO_2 + 6H_2O$
Conditions: burn or spark

Free radical substitution

Reactant: chlorine
Equation: $C_2H_6 + Cl_2 \rightarrow CH_3CH_2Cl + HCl$
Conditions: UV light

Alkenes, e.g. ethene $CH_2{=}CH_2$

Hydrogenation

Reactant: hydrogen
Equation: $CH_2{=}CH_2 + H_2 \rightarrow CH_3CH_3$
Conditions: warm with a nickel (or platinum) catalyst

Electrophilic addition

Reactant: bromine
Equation: $CH_2{=}CH_2 + Br_2 \rightarrow CH_2BrCH_2Br$
Conditions: bromine dissolved in inert solvent

Reactant: bromine water
Equation: $CH_2{=}CH_2 + HOBr \rightarrow CH_2OHCH_2Br$
Conditions: bubble into bromine water

Reactant: hydrogen halides, e.g. HBr
Equation: $CH_2{=}CH_2 + HBr \rightarrow CH_3CH_2Br$
Conditions: mix gases at room temperature

Oxidation

Reactant: aqueous potassium manganate(VII)
Equation: $CH_2{=}CH_2 + H_2O + [O] \rightarrow CH_2OHCH_2OH$
Conditions: bubble into neutral or alkaline solution

Polymerisation

Reactant: ethene
Equation: $nCH_2{=}CH_2 \rightarrow -(CH_2{-}CH_2)-_n$
Conditions: 1000 atm pressure and trace of oxygen

Unit 2

Alcohols, e.g. ethanol C_2H_5OH

Combustion

Reactant: oxygen (air)
Equation: $C_2H_5OH + 3O_2 \rightarrow 2CO_2 + 3H_2O$
Conditions: burn or spark

With sodium

Equation: $C_2H_5OH + Na \rightarrow C_2H_5ONa + ½H_2$
Conditions: mix

Substitution to form halogenoalkanes

Reactant: PCl_5
Equation: $C_2H_5OH + PCl_5 \rightarrow C_2H_5Cl + POCl_3 + HCl$
Conditions: add to dry ethanol

Reactant: HBr
Equation: $C_2H_5OH + HBr \rightarrow C_2H_5Br + H_2O$
Conditions: HBr made in situ from 50% H_2SO_4 and KBr

Reactant: PI_3
Equation: $3C_2H_5OH + PI_3 \rightarrow 3C_2H_5I + H_3PO_3$
Conditions: PI_3 made in situ from moist red phosphorus and iodine

Oxidation to ethanal

Reactant: acidified potassium dichromate(VI)
Equation: $C_2H_5OH + [O] \rightarrow CH_3CHO + H_2O$
Conditions: heat and distil out the aldehyde as it is formed

Oxidation to ethanoic acid

Reactant: acidified potassium dichromate(VI)
Equation: $C_2H_5OH + 2[O] \rightarrow CH_3COOH + H_2O$
Conditions: heat under reflux

Hint

Secondary alcohols are oxidised to ketones; tertiary alcohols are not oxidised.

Halogenoalkanes, e.g. 2-bromopropane $CH_3CHBrCH_3$

Hydrolysis

Reactant: NaOH(aq)
Equation: $CH_3CHBrCH_3 + OH^- \rightarrow CH_3CH(OH)CH_3 + Br^-$
Conditions: warm with aqueous alkali

Elimination

Reactant: KOH in ethanol
Equation: $CH_3CHBrCH_3 + KOH \rightarrow CH_3CH{=}CH_2 + KBr + H_2O$
Conditions: heat under reflux with concentrated KOH in ethanol

With aqueous silver nitrate

Equation: $CH_3CHBrCH_3 + H_2O \rightarrow CH_3CH(OH)CH_3 + HBr$

then $HBr + AgNO_3 \rightarrow AgBr + HNO_3$

Conditions: mix and observe precipitate of silver bromide

With ammonia

Equation: $CH_3CHBrCH_3 + 2NH_3 \rightarrow CH_3CH(NH_2) + NH_4Br$

Conditions: allow to stand with concentrated ammonia

Unit 4

Carbonyl compounds, aldehydes, e.g. CH_3CHO and ketones, e.g. CH_3COCH_3

Oxidation of aldehydes

Reactant: Fehling's or Tollens' solution

Equation: $CH_3CHO + OH^- + [O] \rightarrow CH_3COO^- + H_2O$

Conditions: warm

Observation: Fehling's gives red precipitate, Tollens' gives a silver mirror

Reduction of aldehydes and ketones

Reactant: lithium aluminium hydride

Equation: $CH_3CHO + 2[H] \rightarrow CH_3CH_2OH$

$CH_3COCH_3 + 2[H] \rightarrow CH_3CH(OH)CH_3$

Conditions: dry ether solution; then acidify

Nucleophilic addition to aldehydes and ketones

Reactant: HCN

Equation: $CH_3CHO + HCN \rightarrow CH_3CH(OH)CN$

$CH_3COCH_3 + HCN \rightarrow CH_3C(OH)(CN)CH_3$

Conditions: CN^- ions as catalyst

Test for carbonyl group

Reactant: Brady's reagent (2,4 dinitrophenylhydrazine)

Equation: $CH_3CHO + NH_2NHC_6H_3(NO_2)_2 \rightarrow CH_3CH{=}NNHC_6H_3(NO_2)_2 + H_2O$

Observation: yellow/orange precipitate

Iodoform reaction

Reactant: iodine + alkali

Equation: $CH_3COR + 3I_2 + 4NaOH \rightarrow RCOONa + CHI_3 + 3NaI + 3H_2O$

Conditions: warm gently

Hint

This reaction works with ethanal and with ketones with a CH_3CO group. It also works with ethanol and with secondary alcohols with a $CH_3CH(OH)$ group.

Carboxylic acids, e.g. ethanoic acid, CH_3COOH

Reduction

Reactant: lithium aluminium hydride

Equation: $CH_3COOH + 4[H] \rightarrow CH_3CH_2OH + H_2O$

Conditions: dry ether solution then acidify

Neutralisation

Reactant: sodium hydroxide or sodium carbonate

Equation: $CH_3COOH + NaOH \rightarrow CH_3COONa + H_2O$

$2CH_3COOH + Na_2CO_3 \rightarrow 2CH_3COONa + CO_2 + H_2O$

Conditions: mix or titrate

With PCl_5

Equation: $CH_3COOH + PCl_5 \rightarrow CH_3COCl + POCl_3 + HCl$

Conditions: dry reagents

Esterification

Reactant: ethanol (or other alcohols)

Equation: $CH_3COOH + C_2H_5OH \rightleftharpoons CH_3COOC_2H_5 + H_2O$

Conditions: warm with a catalyst of concentrated sulfuric acid

Acid chlorides, e.g. ethanoyl chloride CH_3COCl

With water

Equation: $CH_3COCl + H_2O \rightarrow CH_3COOH$

Conditions: room temperature

Esterification

Reagent: ethanol (or other alcohols)

Equation: $CH_3COCl + C_2H_5OH \rightarrow CH_3COOC_2H_5 + HCl$

Conditions: dry

With ammonia to form an amide

Equation: $CH_3COCl + 2NH_3 \rightarrow CH_3CONH_2 + NH_4Cl$

Conditions: concentrated ammonia

With amines, e.g. ethylamine to form a substituted amide

Equation: $CH_3COCl + C_2H_5NH_2 \rightarrow CH_3CONHC_2H_5 + HCl$

Conditions: dry

Appendix

Esters, e.g. ethyl ethanoate, $CH_3COOC_2H_5$

Hydrolysis

Reagent: dilute sulfuric acid
Equation: $CH_3COOC_2H_5 + H_2O \rightleftharpoons CH_3COOH + C_2H_5OH$
Conditions: heat under reflux; the acid is the catalyst

Reagent: sodium hydroxide
Equation: $CH_3COOC_2H_5 + NaOH \rightarrow CH_3COONa + C_2H_5OH$
Conditions: heat under reflux in aqueous solution

Transesterification

Reagent: a different alcohol, e.g. methanol
Equation: $CH_3COOC_2H_5 + CH_3OH \rightleftharpoons CH_3COOCH_3 + C_2H_5OH$
Conditions: heat under reflux with a catalyst

Reagent: a different carboxylic acid, e.g. methanoic acid
Equation: $CH_3COOC_2H_5 + HCOOH \rightleftharpoons HCOOC_2H_5 + CH_3COOH$
Conditions: heat under reflux with a catalyst

Polyester formation

Reagent: a diacid (or diacid chloride) plus a diol
Equation: $n HOOC(CH_2)_4COOH + n HO(CH_2)_6OH \rightarrow (-CO(CH_2)_4COO(CH_2)_6O-)_n + n H_2O$

Unit 5

Benzene C_6H_6: electrophilic substitution reactions

Bromination

Reagent: bromine
Equation: $C_6H_6 + Br_2 \rightarrow C_6H_5Br + HBr$
Conditions: liquid bromine and iron to make $FeBr_3$

Nitration

Reagent: concentrated nitric acid
Equation: $C_6H_6 + HNO_3 \rightarrow C_6H_5NO_2 + H_2O$
Conditions: catalyst of concentrated sulfuric at 55°C

Sulfonation

Reagent: sulfur trioxide
Equation: $C_6H_6 + SO_3 \rightarrow C_6H_5SO_3H$
Conditions: fuming sulfuric acid

Friedel–Crafts

Reagent: a halogenoalkane, e.g. chloroethane
Equation: $C_6H_6 + C_2H_5Cl \rightarrow C_6H_5C_2H_5 + HCl$
Conditions: anhydrous $AlCl_3$ catalyst

Reagent: an acid chloride, e.g. ethanoyl chloride
Equation: $C_6H_6 + CH_3COCl \rightarrow C_6H_5COCH_3 + HCl$
Conditions: anhydrous $AlCl_3$ catalyst

Phenol C_6H_5OH

Bromination

Reagent: bromine water
Equation: $C_6H_5OH + 3Br_2 \rightarrow C_6H_2Br_3OH + 3HBr$
Conditions: mix and observe white precipitate

Nitration

Reagent: dilute nitric acid
Equation: $C_6H_5OH + HNO_3 \rightarrow C_6H_4(OH)NO_2 + H_2O$
Conditions: warm gently in aqueous solution

Amines, e.g. ethylamine $C_2H_5NH_2$

Formation of salts

Reagent: hydrochloric acid
Equation: $C_2H_5NH_2 + H^+ \rightarrow C_2H_5NH_3^+$
Conditions: mix aqueous solutions

Complex ion formation

Reagent: hydrated copper(II) ions
Equation: $4C_2H_5NH_2 + [Cu(H_2O)_6]^{2+} \rightarrow [Cu(C_2H_5NH_2)_4(H_2O)_2]^{2+} + 2H_2O$
Conditions: add excess ethylamine

With acid chlorides, e.g. ethanoyl chloride

Equation: $C_2H_5NH_2 + CH_3COCl \rightarrow CH_3CONHC_2H_5 + HCl$
Conditions: room temperature

With halogenoalkanes, e.g. chloroethane

Equation: $C_2H_5NH_2 + C_2H_5Cl \rightarrow (C_2H_5)_2NH + HCl$
Conditions: warm in ethanolic solution

Nitrobenzene $C_6H_5NO_2$

Reduction

Reagent: tin and concentrated hydrochloric acid
Equation: $C_6H_5NO_2 + 6[H] \rightarrow C_6H_5NH_2 + 2H_2O$
Conditions: warm

Phenylamine $C_6H_5NH_2$

Diazotisation and formation of an azo dye

Step 1
Reagent: sodium nitrite and hydrochloric acid
Equation: $C_6H_5NH_2 + NO_2^- + 2H^+ \rightarrow C_6H_5N_2^+ + 2H_2O$
Conditions: keep at 5°C

Step 2

Reagent: phenol

Equation: $C_6H_5N_2^+ + C_6H_5OH + OH^- \rightarrow$
$C_6H_5\text{-N=N-}C_6H_4OH + H_2O$

Conditions: aqueous alkali

Formation of polyamides

Reagents: a diacid chloride and a diamine

Equation: $n\text{ClOC(CH}_2)_4\text{COCl} + n\text{NH}_2(\text{CH}_2)_6\text{NH}_2 \rightarrow$
$(-\text{CO(CH}_2)_4\text{CONH(CH}_2)_6\text{NH}-)_n + n\text{HCl}$

Conditions: mix the diacid chloride with an aqueous solution of the diamine

A

B

C

The periodic table

Key:

Relative atomic mass
Atomic symbol
name
Atomic (proton) number

Period	1	2	Group										3	4	5	6	7	0
1	1.0 H hydrogen 1																	4.0 He helium 2
2	6.9 Li lithium 3	9.0 Be beryllium 4											10.8 B boron 5	12.0 C carbon 6	14.0 N nitrogen 7	16.0 O oxygen 8	19.0 F fluorine 9	20.2 Ne neon 10
3	23.0 Na sodium 11	24.3 Mg magnesium 12											27.0 Al aluminium 13	28.1 Si silicon 14	31.0 P phosphorus 15	32.1 S sulfur 16	35.5 Cl chlorine 17	39.9 Ar argon 18
4	39.1 K potassium 19	40.1 Ca calcium 20	45.0 Sc scandium 21	47.9 Ti titanium 22	50.9 V vanadium 23	52.0 Cr chromium 24	54.9 Mn manganese 25	55.8 Fe iron 26	58.9 Co cobalt 27	58.7 Ni nickel 28	63.5 Cu copper 29	65.4 Zn zinc 30	69.7 Ga gallium 31	72.6 Ge germanium 32	74.9 As arsenic 33	79.0 Se selenium 34	79.9 Br bromine 35	83.8 Kr krypton 36
5	85.5 Rb rubidium 37	87.6 Sr strontium 38	88.9 Y yttrium 39	91.2 Zr zirconium 40	92.9 Nb niobium 41	95.9 Mo molybdenum 42	[98] Tc technetium 43	101.1 Ru ruthenium 44	102.9 Rh rhodium 45	106.4 Pd palladium 46	107.9 Ag silver 47	112.4 Cd cadmium 48	114.8 In indium 49	118.7 Sn tin 50	121.8 Sb antimony 51	127.6 Te tellurium 52	126.9 I iodine 53	131.3 Xe xenon 54
6	132.9 Cs caesium 55	137.3 Ba barium 56	138.9 La lathanum 57	178.5 Hf hafnium 72	180.9 Ta tantalum 73	183.8 W tungsten 74	186.2 Re rhenium 75	190.2 Os osmium 76	192.2 Ir iridium 77	195.1 Pt platinum 78	197.0 Au gold 79	200.6 Hg mercury 80	204.4 Tl thallium 81	207.2 Pb lead 82	209.0 Bi bismuth 83	[209] Po polonium 84	[210] At astatine 85	[222] Rn radon 86
7	[223] Fr francium 87	[226] Ra radium 88	[227] Ac actinium 89	[261] Rf rutherfordium 104	[262] Db dubnium 105	[266] Sg seaborgium 106	[264] Bh bohrium 107	[277] Hs hassium 108	[268] Mt meitnerium 109	[271] Ds darmstadtium 110	[272] Rg roentgenium 111	Elements with atomic numbers 112–116 have been reported but not fully authenticated						

140.1 Ce cerium 58	140.9 Pr praseodymium 59	144.2 Nd neodymium 60	144.9 Pm promethium 61	150.4 Sm samarium 62	152.0 Eu europium 63	157.2 Gd gadolinium 64	158.9 Tb terbium 65	162.5 Dy dysprosium 66	164.9 Ho holmium 67	167.3 Er erbium 68	168.9 Tm thulium 69	173.0 Yb ytterbium 70	175.0 Lu lutetium 71
232 Th thorium 90	[231] Pa protactinium 91	238.1 U uranium 92	[237] Np neptunium 93	[242] Pu plutonium 94	[243] Am americium 95	[247] Cm curium 96	[245] Bk berkelium 97	[251] Cf californium 98	[254] Es einsteinium 99	[253] Fm fermium 100	[256] Md mendelevium 101	[254] No nodelium 102	[257] Lr lawrencium 103